ELECTRICAL MAINTENANCE

THE INSTITUTION OF ELECTRICAL ENGINEERS

Published by: The Institution of Electrical Engineers
Savoy Place, LONDON
United Kingdom. WC2R 0BL

Copies may be obtained from:
The Institution of Electrical Engineers
PO Box 96, STEVENAGE
United Kingdom. SG1 2SD.

Tel: 01438 767 328
Fax: 01438 742 792
Email: sales@iee.org.uk
http;//www.iee.org.uk/publish/

While the author and the publisher believe that the information and guidance given in this work is correct, all parties must rely upon their own skill and judgement when making use of it. Neither the author nor the publisher assume any liability to anyone for any loss or damage caused by any error or omission in the work, whether such error or omission is the result of negligence or any other cause. Any and all such liability is disclaimed.

ISBN 085296 769 1

ELECTRICAL MAINTENANCE

Contents

Co-operating Organisations

The Institution of Electrical Engineers acknowledge the contribution made by the following representatives of organisations in the preparation of this Publication.

Association of Consulting Engineers
P Lawson Smith

BEAMA
British Electrical Equipment Manufacturers Association
M Mullins

Department of the Environment, Transport and the Regions
E N King

Department of Trade and Industry
D Dolbey-Jones

Electrical Contractors Association
P J Buckle

Electrical Contractors Association of Scotland
D N McGuiness

Gambica Association Ltd
K Morriss

Health and Safety Executive
R T R Pilling

Institution of Electrical Engineers
N C Friswell

Institution of Lighting Engineers
G Pritchard

Lighting Association
K R Kearney

Lighting Industry Federation
B Pratley

National Inspection Council for Electrical Installation Contracting
J T Bradley

Safety Assessment Federation Limited
A Daley

P Cook   Editor

D W M Latimer   Publications Committee Chairman

ELECTRICAL MAINTENANCE

Chapter 1   THE NEED FOR MAINTENANCE

1.1   General

Electrical maintenance is carried out for four basic reasons :

1)   To prevent danger
2)   To reduce unit cost
3)   To keep a facility in operation (reliability)
4)   To prevent pollution of the environment

The need to prevent danger is enforced by much legislation. Cost and reliability are matters to be decided probably on managerial criteria only. Protection of the environment is supported by legislation.

1.2   Prevention of danger

Apart from the common law general duty of care of everyone for his neighbour there is specific legislation with respect to safety, the most fundamental being the Health and Safety at Work etc. Act 1974. This Act empowers the Secretary of State to make regulations, generally referred to as Health and Safety Regulations. A number of these Health and Safety Regulations are discussed where they are relevant to electrical maintenance. The Health and Safety at Work etc. Act empowers the Health and Safety Commission to approve and issue codes of practice. These codes of practice may be prepared by the Commission or by others. The consent of the Secretary of State is required before the Commission approves a code of practice. The failure on the part of any person to observe a provision of an approved code of practice does not of itself render him liable to any civil or criminal proceedings, but the breach of a provision of a code of practice could be used as evidence of non-compliance with a regulation. The approved codes of practice issued by the Health and Safety Commission consequently have a special place in the Health and Safety at Work legislation, those concerned with electrical maintenance are summarised in Appendix A with the relevant legislation.

Health and Safety at Work etc. Act 1974

The Health and Safety at Work Act 1974 is comprehensive, concerned with health, safety and welfare at work, the control of dangerous substances and certain emissions into the atmosphere.

Sections 2, 3 and 4 of the Health and Safety at Work Act 1974 put a duty of care upon both the employer, the employee and the self employed to ensure the health, safety and welfare at work of all persons, employees and others using the work premises.

Section 6 imposes a duty on any person who designs or manufactures any article for use at work to take such steps as are necessary to ensure that there will be available, in connection with the use of the article at work, adequate information on the use for which the article has been designed and tested. It also requires that advice be given on any conditions necessary to ensure that, when put to use, it will be safe, and without risks to health. This implies a general duty that persons supplying electrical installations and electrical equipment must ensure that not only the equipment they supply is safe and adequately tested, but that adequate information is provided for it to be maintained in a safe condition.

Of the many health and safety regulations issued under the Health and Safety at Work Act, perhaps the most relevant for electrical installations and electrical equipment are the Electricity at Work Regulations 1989.

The Electricity at Work Regulations 1989

The Electricity at Work Regulations (SI 1989 No 635) impose duties on every employer, every employee and every self employed person to ensure that the safety requirements of the regulations are met. The requirements of regulation 4 are:

1) All systems shall at all times be of such construction as to prevent, so far as is reasonably practicable, danger.
2) As may be necessary to prevent danger, all systems shall be maintained so as to prevent, so far as is reasonably practicable, such danger.
3) Every work activity, including operation, use and maintenance of a system and work near a system, shall be carried out in such a manner as not to give rise, so far as is reasonably practicable, to danger.
4) Any equipment provided under these Regulations for the purpose of protecting persons at work on or near electrical equipment shall be suitable for the use for which it is provided, maintained in a condition suitable for that use, and be properly used.

In effect, all electrical systems must be designed, installed, maintained and used so as to prevent danger. The term "system" means electrical system and includes all the electrical equipment that is, or may be, connected to a common source of electrical energy. This includes the fixed installation and all equipment that may be supplied from it and as such also embraces equipment supplied via a plug and socket.

It is principally with the electrical installation and fixed electrical equipment that this publication is concerned. For the maintenance of portable equipment, appliances, business and office equipment, the IEE has published a code of practice for In-Service Inspection and Testing of Electrical Equipment.

The regulations include many requirements specific to the nature of the installation. However, the Memorandum of Guidance on the Electricity at Work Regulations (see Appendix A) advises that the IEE Wiring Regulations (BS 7671): Requirements for Electrical Installations, is a code of practice which is widely recognised and accepted in the UK, and compliance with it is likely to achieve compliance with the relevant aspects of the 1989 regulations.

The scope of the Electricity at Work Regulations is much wider than BS 7671 in that they require:

installations to be constructed so as to be safe;
installations to be maintained so as to be safe;
associated work to be carried out safely;
work equipment provided to be suitable for the purpose.

The Electricity at Work Regulations are also concerned with maintenance of the installation, training and competency of staff, good working practices and suitable equipment. It is to deal with these matters that this publication has been written.

Management of Health and Safety at Work Regulations 1992

Approved Code of Practice L21

The prime requirement of the Management of Health and Safety at Work Regulations is that :

Every employer shall make a suitable and sufficient assessment of:

a)   The risks to the health and safety of his employees to which they are exposed while they are at work and

b)   the risks to the health and safety of persons not in his employment arising out of or in connection with the conduct by him of his undertaking. (Regulation 3).

The Management of Health and Safety at Work Regulations apply to all risks to health and safety of persons whereas the Electricity at Work Regulations deal with electrical systems and work on, associated with or near such systems. The Management Regulations introduce the concept of risk assessment in the broadest sense. The level of detail required within the risk assessment is to be proportionate to the hazard. Significant findings are required to be recorded, together with the control measures taken.

These regulations have requirements for health and safety training. Work entrusted to employees should take into account their capabilities and adequate health and safety training is required to be provided (Regulation 11). Regulation 16 of the Electricity at Work Regulations requires that no person shall be engaged in any work activity where technical knowledge or experience is necessary to prevent danger or, where appropriate injury, unless they possess such knowledge or experience or are under a degree of supervision as may be appropriate having regard to the nature of the work.

As well as imposing duties upon the employer, the regulations impose duties on employees. Every employee is required to use machinery, equipment, etc. provided to him by his employer in accordance with any training he has received and any instructions provided to him (Regulation 12). There is also a duty upon every employee to inform his employer where he has concerns regarding the safety of fellow employees, both with respect to equipment provided and training. This is a particularly relevant requirement for those with the general responsibilities for maintenance, including those carrying out the maintenance work. If any person has concerns regarding his ability to carry out work he must bring this to the attention of his employer.

Workplace (Health, Safety and Welfare) Regulations 1992

Approved Code of Practice L24

The Workplace (Health, Safety and Welfare) Regulations 1992 require that every employer shall ensure that the workplace equipment, devices and systems are maintained. This includes keeping the equipment devices and systems in an efficient state, in efficient working order, and in good repair. Where appropriate, the equipment, devices and systems shall be subject to a suitable system of maintenance.

The scope of the Workplace (Health, Safety and Welfare) Regulations is somewhat different to the Electricity at Work Regulations. The Electricity at Work Regulations are basically concerned with the maintenance of the electrical installation in a safe condition and with safe working upon it. They do not deal with the consequences of maloperation of the electrical system. However, the Workplace Regulations are concerned with the consequences of equipment and system failures. Whilst a malfunctioning emergency lighting system may not be an electrical hazard, there is of course a potential hazard if there is no emergency lighting. These regulations impose maintenance regimes upon such systems as emergency lighting, fire alarms, powered doors, escalators and moving walkways that have electrical power supplies. The regulations are much broader than electrical systems and also include fencing, equipment used for window cleaning, devices to limit the opening of windows etc. The approved code of practice to the Workplace Regulations states that the maintenance of work electrical equipment and electrical systems is also addressed in other regulations. Electrical systems are clearly well addressed in the Electricity at Work Regulations and the maintenance of work equipment in the Provision and Use of Work Equipment Regulations 1992.

Construction (Design and Management) Regulations 1994

Approved Code of Practice L54

The Construction (Design and Management) Regulations 1994 generally apply to construction work (Regulation 3). The basic requirement is that design and construction must take account of health and safety aspects in the construction phase of the work and during any subsequent maintenance. They also have requirements regarding the cleaning, maintaining and repairing of the construction at any time, including after construction is completed, i.e. during use and during demolition of the construction.

There is a requirement to provide reasonably foreseeable information necessary for the health and safety of persons who will carry out maintenance, repairs and cleaning in the future. Consequently, persons with such responsibilities for a building should request access to the health and safety file prepared for the construction to see if there are any particular problems associated with maintenance and repair, including electrical matters.

Provision and Use of Work Equipment Regulations 1998

HSE Guidance on Regulations L22

The Provision and Use of Work Equipment Regulations require that work equipment including installations, is so constructed (or adapted) so as to be suitable for the purpose for which it is provided. Equipment must be inspected after installation and before use, at suitable intervals and after exceptional events e.g. circuit-breaker operation under fault conditions. Potentially dangerous machinery must be guarded. Where necessary, logs maintained and adequate training give to operations.

Personal Protective Equipment at Work Regulations 1992

HSE Guidance on Regulations L25

The Personal Protective Equipment at Work Regulations require every employer to ensure that suitable personal protective equipment is provided to his employees, as may be necessary. The equipment must take account of the risks, the environmental conditions, ergonomic requirements, the state of the health of the person or persons, it must fit and it must comply with any appropriate provisions or standards.

The regulations require that an assessment be made to ensure that the protective equipment is suitable. Purchasers should be looking for approval body and CE marking to demonstrate compliance with appropriate standards. The relevance of the approval mark to the use must be confirmed. The Notes of Guidance published by the HSE make the important note that the initial selection is only a first stage in a continuing programme concerned with the proper selection, maintenance and use of the protective equipment. Training and supervision of users must be appropriate to the equipment. The employer is required to ensure that the employee has sufficient information, instruction and training to use the equipment provided and knows the limits of the protection provided by the equipment and any supplementary action that might be necessary in particular circumstances. For example, in the selection of safety goggles, the employee must be made aware as to what the goggles are protecting against, as of course one pair may not be suitable for all the potential risks.

There is a requirement upon employees to use protective equipment provided in accordance with training and instruction received.

Employees are required to report the loss of such equipment. The requirements of this legislation does mean that proper records should be kept of protective equipment and a suitable procedure be set up for checking that employees still have the equipment necessary and that it is in good order. This does not in any way reduce the duty of the employee to advise the employer of any defects or deficiencies in the equipment or the training that he or she has received.

Manual Handling Operations Regulations 1992

HSE Guidance on Regulations L23

Employers are required so far as is reasonably practicable to avoid the need for employees to undertake any manual handling operations which involve the risk of their being injured. Where this is not practicable, employers are obliged to make a suitable and sufficient assessment of the manual handling operation and take appropriate steps to reduce the risk of injury to the lowest level reasonably practicable.

The manual handling guidance L23 published by the Health and Safety Executive provides general guidance on manual handling operations and includes examples of assessment checklists, which are reproduced in Figures 1.1a and 1.1b.

Health and Safety (Display Screen Equipment) Regulations 1992

HSE Guidance Note L26

The Health and Safety (Display Screen Equipment) Regulations 1992 place a requirement upon employers to ensure that work stations (desks with computer screens and keyboards) meet the requirements laid down in schedules to the regulations.

An employer is required at the request of an employee to provide an appropriate eyesight test for users of workstations or persons who are about to become users. There is a further requirement that the employer provides adequate training and information. These regulations are discussed further in section 10.3.

The Health and Safety (Signs and Signals) Regulations 1996

HSE Guidance Note L64

These Regulations are discussed in Chapter 11.

The Fire Precautions (Work Place) Regulations 1997

The Home Office and Scottish Office publish a guidance document "Fire precautions in the work place", ISBN 0 1134 116 9 providing information for employers about the Fire Precautions (Work Place) Regulations 1997.

The Fire Precautions (Work Place) Regulations require every employer where necessary to safeguard the safety of employees in the case of fire. They are required to equip the work place with:

Fire fighting equipment
Fire detectors and alarms.

It is a requirement that any non-automatic fire fighting equipment shall be easily accessible, simple to use and indicated by signs. The employer is required to take measures for fire fighting appropriate for the size and nature of the activity, number of employees etc., to nominate a person to implement these measures and to ensure that the arrangements are complied with.

Where necessary to safeguard the safety of employees arrangements must be made to ensure :

Routes to exits and the exits themselves are kept clear
Emergency routes lead to a place of safety
It is possible to evacuate the workforce quickly and safely
Doors open in the direction of escape (sliding and revolving doors should not be the means of escape)
Emergency doors are not locked or fastened
Emergency routes and exits are indicated by signs
Emergency routes and exit signs are illuminated.

1.3   Cost and reliability

Balancing the cost of maintenance against cost of breakdown generally involves managerial decisions that may be divorced from safety considerations. These matters are discussed in Chapter 2.

1.4   The environment

This publication deals with electrical maintenance. However, many electrical maintenance activities produce waste (oil, old lamps, etc.) the impact of which upon the environment needs consideration. This is discussed in Chapter 16 together with relevant legislation.

Figure 1.1a Example of an assessment checklist

Manual handling of loads

EXAMPLE OF AN ASSESSMENT CHECKLIST

Note : This checklist may be copied freely. It will remind you of the main points to think about while you:

- consider the risk of injury from manual handling operations
- identify steps that can remove or reduce the risk
- decide your priorities for action.

SUMMARY OF ASSESSMENT	Overall priority for remedial action: Nil/Low/Med/High*
Operations covered by this assessment :	Remedial action to be taken: ....
Locations :	....
Personnel involved :	....
Date of assessment :	Date by which action is to be taken : ....
	Date for reassessment : ....
	Assessor's name: ....
	Signature: ....

* circle as appropriate

Section A - Preliminary:

Q1   **Do the operations involve a significant risk of injury?**   Yes/No*

   If 'Yes' go to Q2. If 'No' the assessment need go no further.

   If in doubt answer 'Yes'. You may find the guidelines in Appendix 1 helpful

Q2   **Can the operations be avoided/mechanised/automated at reasonable cost?**   Yes/No*

   If 'No' go to Q3. If 'Yes' proceed and then check that the result is satisfactory

Q3   **Are the operations clearly within the guidelines in Appendix 1?**   Yes/No*

   If 'No' go to Section B. If 'Yes' you may go straight to Section C if you wish.

Section C - Overall assessment of risk:

Q   **What is your overall assessment of the risk of injury?**   Insignificant/Low/Med/High*

   If not 'Insignificant' go to Section D. If 'Insignificant' the assessment need go no further

Section D - Remedial action:

Q   **What remedial steps should be taken, in order of priority?**

   I..

   ii...

   iii..

   iv...

   v..

And finally:

- complete the SUMMARY above
- compare it with your other manual handling assessments
- decide your priorities for action
- **TAKE ACTION.............AND CHECK THAT IT HAS THE DESIRED EFFECT**

14

Figure 1.1b

Section B - More detailed assessment, where necessary:				
Questions to consider: (if the answer to a question is 'Yes' place a tick against it then consider the level of risk)		**Level of risk:** (Tick as appropriate)			**Possible remedial action:** (Make rough notes in this column in preparation for completing Section D)
	Yes	**Low**	**Med**	**High**	
The tasks - do they involve:					
◆ holding loads away from trunk?					
◆ twisting?					
◆ stooping?					
◆ reaching upwards?					
◆ large vertical movement?					
◆ long carrying distances?					
◆ strenuous pushing or pulling?					
◆ unpredictable movement of loads?					
◆ repetitive handling?					
◆ insufficient rest or recovery?					
◆ a workrate imposed by a process?					
The loads - are they:					
◆ heavy?					
◆ bulky/unwieldy?					
◆ difficult to grasp?					
◆ unstable/unpredictable?					
◆ intrinsically harmful (e.g. sharp/hot)?					
The working environment - are there:					
constraints on posture?					
poor floors?					
◆ variations in levels?					
◆ hot/cold/humid conditions?					
◆ strong air movements?					
◆ poor lighting conditions?					
Individual capability - does the job:					
◆ require unusual capability?					
◆ hazard those with a health problem?					
◆ hazard those who are pregnant?					
◆ call for special information/training?					
Other factors -					
Is movement or posture hindered by clothing or personal protective equipment?					
Deciding the level of risk will inevitably call for judgement. The guidelines in Appendix 1 may provide a useful yardstick. **When you have completed Section B go to Section C.**				

Chapter 2   MAINTENANCE MANAGEMENT

2.1   Maintenance categories

There are three general categories of maintenance:

1)   Breakdown;
2)   Preventive;
3)   Condition monitored.

Breakdown Maintenance

Breakdown maintenance is the simplest approach. Equipment and systems are repaired or replaced when they cease to work. This approach can be followed where breakdown or failure to work does not result in danger and the consequences of the failure or breakdown are otherwise acceptable. This approach would be reasonable and is adopted for the failure of lamps in domestic premises.

Preventive maintenance

Preventive maintenance is carried out before breakdown or failure occurs. This will normally mean that it is planned and carried out at specified intervals, although it may be initiated by a signal or feedback of some description. Two examples of preventive maintenance would be the routine changing of street lighting lamps and the changing of the engine oil of a generator prime mover.

Condition monitored maintenance

Condition monitored maintenance requires the measurement of certain parameters of the equipment, such as vibration or temperature. At pre-set levels, alarms are initiated allowing the equipment to be shut down and maintenance carried out. This is appropriate for such equipment as large motors where the cost of breakdown is high and/or where shutdown is to be limited to as few occasions as possible.

2.2   Keeping of records

Much legislation simply requires that either maintenance be carried out or, more commonly, as in the Electricity at Work Regulations, that equipment be maintained so as to prevent, so far as is reasonably practicable, danger. There is generally no specific requirement to keep records. However, in most of the guidance issued by the Health and Safety Executive, including the Memorandum of Guidance on the Electricity at Work Regulations 1989, the advice is given that records of maintenance including test results, preferably kept throughout the working life of an electrical system, will enable the condition of the equipment and the effectiveness of the maintenance policies to be monitored. Additionally, without effective monitoring, duty holders cannot demonstrate that the requirement for maintenance has been met.

2.3   Inspection and testing

Rarely in legislation is there a specific requirement for inspection and testing. The requirement is for maintenance, either at specified intervals or to ensure safety, as in the Electricity at Work Regulations. The purpose of inspection and testing is to determine if any maintenance, including repairs, is required. The keeping of inspection and test results will also demonstrate that, as far as the inspector and tester could reasonably determine, maintenance was or was not required at the time of the inspection and test. The records will also enable slow deterioration or step changes in the condition of equipment to be identified.

2.4   Maintenance strategies

Maintenance is basically carried out for three reasons:

1)   Safety;
2)   Protection of the environment;
3)   Minimising business costs and maximising income.

Safety

Maintenance for safety may be carried out to meet the following general requirements:

1)   Common law requirements;
2)   Explicit legal requirements;
3)   Implied legal requirements.

The common law implies a general duty of care to other persons and their livestock, property, etc.

Explicit legal requirements are those where there are particular requirements in the legislation for maintenance. These are summarised in Table 2A. This type of maintenance requirement is not common. More frequently, as in the Electricity at Work Regulations, the requirement is phrased as follows:

"As may be necessary to prevent danger, all systems shall be maintained so as to prevent, so far as is reasonably practicable, such danger." (Regulation 4(2)).

There is no requirement here to carry out a maintenance activity as such; the requirement is simply to maintain the system (including equipment) so as to prevent danger. This is an implied maintenance requirement as generally equipment cannot be maintained in a safe condition without actually being maintained. Normally it is necessary to inspect and/or test a system to determine if maintenance (including repairs) is necessary. It is also likely to be necessary to monitor the effectiveness of any maintenance procedures. The legislation with implied requirements for maintenance is listed in Table 2B.

Protection of the environment

Maintenance may be required to be carried out, not simply to protect people's health and safety, but also to protect the environment. This may not be cost effective, but it may be seen as a general duty of care or it may be a legislative requirement as required by the Clean Air Act or the Environmental Protection Act and associated Regulations.

Minimising costs and maximising business income

Maintenance carried out to reduce the cost of an enterprise would include action taken to reduce or avoid:

1)   The cost of failure of plant or equipment - repair costs;
2)   The cost of loss of production - revenue cost;
3)   The cost of loss of service - revenue and goodwill.

Repair cost

Decisions on the approach to be taken are rarely simple. For example, all the lamps of a street lighting installation will need to be replaced at some time or the street will end up in darkness. The decision as to whether it is cost effective to replace the lamps routinely (preventive maintenance) or when they fail (breakdown) will need to take into account the reduction in light output (if significant), the effect on traffic of frequent disturbances and the cost of attending to replace lamps. It might be decided that breakdown maintenance was appropriate for a B road, whilst preventive lamp replacement was necessary for a motorway.

A balance needs to be achieved between the cost of the maintenance activity and the cost of the equipment.

The cost of maintaining a large motor might well be small compared with the cost of replacement, whereas the cost of replacing a single tungsten filament lamp will far outweigh the cost of the lamp. The lamp may be maintained on a breakdown basis, the motor on a routine basis.

Loss of production

In many situations the cost of the failed piece of equipment is insignificant compared with the cost of loss of output or production. In these circumstances breakdown maintenance is unlikely to be appropriate.

Breakdown of service

Customer goodwill is difficult to estimate financially, but should be considered, when determining maintenance regimes. The additional costs of early replacement, or even frequent maintenance, can be justified by customer goodwill.

Table 2A   Specific maintenance requirements

Certain types of installation require formal inspection by a "Competent Person" as an explicit statutory requirement. The following table gives examples.

TYPE OF INSTALLATION	MINIMUM FREQUENCY OF INSPECTION	LEGISLATION
Lifts such as electric and hydraulic passenger and goods lifts.	Every 6 months	Requirement of the following legislation: The Factories Act 1961. The Offices, Shops and Railway Premises (Lift and Hoist) Regulations 1968. The Lifting Plant and Equipment (Records of Test and Examination etc.) Regulations 1992.
Lifting appliances such as runway beams, manually and electrically operated hoist blocks etc.	Every 14 months	Requirement of the Factories Act 1961 and the Construction (Lifting Operations) Regulations 1961.
Lifting tackle, chain and wire rope slings, eye bolts, shackles etc.	Every 6 months	Requirement of the Factories Act 1961.
Hoists such as scissors lifts, platform lifts, dock levellers, builder's hoists. (For buildings under repair, a joint responsibility with the contractor).	Every 6 months	Requirement of the Offices, Shops and Railway Premises (Lift and Hoist) Regulations 1968 and the Construction (Lifting Operations) Regulations 1961.
Cranes such as jib cranes, overhead travelling cranes, mobile cranes etc.	Every 14 months and after every substantial alteration or repair.	Requirement under both the Factories Act 1961 and the construction (Lifting Operations) Regulations 1961.
Suspended access equipment, e.g. window cleaning and façade maintenance equipment.	Not stated	Covered by the Construction (Working Places) Regulations 1966 and the Construction (Lifting Operations) Regulations 1961.
Pressure vessels and systems (excluding boilers) such as air receivers and steam receivers.	See COP 37 and 30.	Requirements of the Pressure Systems and Transportable Gas Containers Regulations 1989.
Pressurised boiler plant and systems, including steam.	Every 14 months or as specified in a Written Scheme of Examination	The inspection of steam boilers is also covered by the Examination of Steam Boilers Regulations 1964.
Electrical installations in cinemas.	Annually	An explicit requirement of the Cinematography Act 1909.

Note:

In addition to the above, there may be requirements for periodic inspection of certain installations in local bylaws.

Table 2B   Implied maintenance requirements

The following table gives an indication of the minimum maintenance that must be carried out in order to meet explicit or implied statutory requirements.

SYSTEM	TASK	SPECIFIC OR RECOMMENDED MINIMUM FREQUENCY	REASON
Electrical installations	1. Carry out routine checks. 2. Inspect and as necessary test.	See Tables 4A and 4B.	To comply with the Electricity at Work Regulations 1989.
Inspection and testing of electrical equipment including hand-held, portable, stationary and fixed.	Inspection and as necessary testing of all electrical equipment.	See Table 12A	To comply with the Electricity at Work Regulations 1989.
Maintenance of emergency lighting systems (See Chapter 5)	1. Inspect the emergency lighting installation, and test each self-contained luminaire, internally illuminated exit sign, central battery installation and emergency lighting standby generator.	Monthly	To comply with the Fire Precautions Act 1971 and, for certain dwellings, the Building Regulations 1991. Compliance with BS 5266 is deemed to satisfy. Workplace (Health, Safety and Welfare) Regulations 1992.
	2. Inspect each self-contained luminaire and internally illuminated exit sign, central battery installation and emergency lighting standby generator and do extended tests.	6 monthly	
	3. Thoroughly inspect and test the emergency lighting installation, central battery and emergency lighting standby generator.	3 yearly	
	4. Do full duration test on each self-contained luminaire and internally illuminated exit sign.	After 3 years, then annually	

SYSTEM	TASK	SPECIFIC OR RECOMMENDED MINIMUM FREQUENCY	REASON
Maintenance of fire detection and alarm systems (See Chapter 6)	1. Carry out reporting procedure and attend to outstanding faults.	Every scheduled visit	To comply with the Fire Precautions Act 1971 and the Electricity at Work Regulations 1989. Compliance with BS 5839, manufacturer's/ supplier's instructions and BS 7671 is deemed to satisfy.
	2. Thoroughly inspect and test power supplies, batteries and chargers, controls and indicators etc.	3 monthly	
	3. Inspect and test the alarm devices.	3 monthly	
	4. Check for changes in building structure and usage, which prejudice the correct operation of the system.	3 monthly	Workplace (Health and Safety and Welfare)
	5. Thoroughly inspect and test all call points and response of all detectors/sensors to appropriate fire product.	Annually, or one quarter of the points 3 monthly	
Maintenance of overhead travelling cranes and runways	1. Check all components such as gearing, shafting, bearings, brakes, wire ropes, anchorages, rope drums, pulleys, hooks, rails, wheels etc.	3 monthly	To comply with Section 27 of the Factories Act 1961.
	2. Check oil in gear cases and bearings: top up as necessary.	3 monthly	
	3. Check generators, control panels, radio/infra-red transmitters, conductor rails, festoon cables, limit switches, proximity devices, audible warning devices etc.	3 monthly	
	4. Check rail tracks, points and structures, runway track changeover switches.	6 monthly	

SYSTEM	TASK	SPECIFIC OR RECOMMENDED MINIMUM FREQUENCY	REASON
Heating and domestic hot water systems	**Storage vessels:** 1. Check stored temperatures under no draw-off conditions. 2. Check outflow water temperatures under normal draw-off conditions. 3. Test water samples from drain for microbial activity. 4. Inspect internal conditions.	Subject to a Risk Assessment. 6 Monthly for buildings with spray outlets and annually for other buildings.	To comply with the Control of Substances Hazardous to Health (COSHH) Regulations 1994. Compliance with the HSC Approved Code of Practice (The Prevention or Control of Legionellosis) and the guidance in HSE Guidance Booklet HS(G)70 (The Control of Legionellosis including Legionnaires' Disease) is deemed to satisfy. Medical buildings should, in addition, comply with the DHSS Code of Practice "The Control of Legionellae in Health Care Premises.
	Hot water service systems: 5. Check branch, sub-branch and main return water temperatures under normal and no draw-off conditions. 6. Test samples from outlets for microbial activity. 7. Clean and disinfect shower heads.	Subject to a Risk Assessment, 6 monthly for buildings with spray outlets and annually for other buildings.	

SYSTEM	TASK	SPECIFIC OR RECOMMENDED MINIMUM FREQUENCY	REASON
Operation of boiler plant with rated output above 150 kW	1. Observe the furnace and chimney top for smoke.	Daily	Compliance with the Clean Air Act and the HSE Guidance Note PM5 for automatically controlled steam and hot water boilers.
	2. Steam boilers only: (1) Blow down water level gauges (2) Blow down water level controls having separately mounted float chambers (3) Check direct mounted water level controls	Daily	
	3. Check water treatment programmes:		
	3A Steam boilers only: (1) Test make-up and feed water (2) Record make-up consumption (3) Test boiler water quality (4) Adjust rate of dosing as necessary (5) Adjust frequency of boiler slowdown as necessary	Daily	
	3B Hot water boilers only: (1) Test make-up water (2) Record make-up consumption (3) Test system water quality (4) Adjust rate of dosing as necessary	Weekly	
	4. Test water level controls and alarms:		
	4A Steam boilers only: Test water level controls and alarms - (1) Low water (2) Second low water (3) High water	Weekly	

SYSTEM	TASK	SPECIFIC OR RECOMMENDED MINIMUM FREQUENCY	REASON
	4B Hot water systems only: (1) Drain down water level controls of pressurisation vessels having separately mounted float chambers (2) Check direct mounted (internal) water level controls of pressurisation vessels	Weekly	
	5. Test fuel feed cut-off:	Monthly	
	5A. Oil-fired boilers only - test the operation of the fire valve and examine the fusible link, cable and pulleys etc. for free operation.		

SYSTEM	TASK	SPECIFIC OR RECOMMENDED MINIMUM FREQUENCY	REASON
Cold water systems	1. Bacteriological analysis of water taken from stored drinking water cisterns containing more than 1000 litres. 2. Chemical analysis of water taken from stored drinking water cisterns containing more than 1000 litres. 3. Inspection.	6 monthly Annually Annually	To comply with the Offices, Shops and Railway Premises Act and the Health and Safety at Work etc. Act 1974. To ensure continued compliance with Water Bylaws.
Operation of a refrigeration installation	Check water treatment programme for evaporative cooling systems: 1. Test system water quality 2. Record make-up water consumption and test quality	Weekly	Compliance with the HSC Approved Code of Practice (The Prevention or Control of Legionellosis) and the guidance in HSE Guidance Booklet HS (G) 70 (The Control of Legionellosis including Legionnaires' Disease) is deemed to satisfy.
Maintenance of a refrigeration installation.	1. Seasonally clean and/or disinfect reservoirs and pipework of evaporative cooling systems	6 monthly, or upon restarting the systems following a period of non-use exceeding one month.	Compliance with the HSC Approved Code of Practice (The Prevention or Control of Legionellosis) and the guidance in HSE Guidance Booklet HS(G)70 (The Control of Legionellosis including Legionnaires' Disease) is deemed to satisfy.
Operation of an air conditioning installation	Check water treatment programme for humidifiers/spray washers: 1. Test system water quality 2. Record make-up water consumption and test quality	Weekly	Compliance with the HSC Approved Code of Practice (The Prevention of Control of Legionellosis) and the guidance in HSE Guidance Booklet HS(G) 70 (The Control of Legionellosis including Legionnaires' Disease) is deemed to satisfy.

SYSTEM	TASK	SPECIFIC OR RECOMMENDED MINIMUM FREQUENCY	REASON
Maintenance of an air conditioning installation	1. Seasonally clean and/or disinfect permanently or intermittently wetted surfaces within the air handling system.	6 monthly	Compliance with the HSC Approved Code of Practice (The Prevention or Control of Legionellosis) and the guidance in HSE Guidance Booklet HS(G)70 (The Control of Legionellosis including Legionnaires' Disease) is deemed to satisfy.

SYSTEM	TASK	SPECIFIC OR RECOMMENDED MINIMUM FREQUENCY	REASON
Operation and maintenance of water treatment plant for swimming pools	1. Test pool water pH and adjust rate of dosing 2. Test for free combined and total chlorine and adjust recirculation rate as necessary 3. Examine pool water for suitability for bathing 4. Check operating condition of disinfection, pH correction and control system for the service water 5. Check ozone generation equipment.	3 times Daily 4 times Daily Monthly Daily Twice Daily	To comply with the Control of Substances Hazardous to Health (COSHH) Regulations 1994. Compliance with the HSE Guidance Note EH38 and the British Effluent and Water Association (BEWA) Code of Practice is deemed to satisfy.
Asbestos based materials - safety procedure	1. Inspect asbestos lagging and materials containing asbestos which form part of M&E installations, except where the lagging or materials are installed on pipes in duct. 2. Inspect asbestos based material installed on pipes in duct.	Annually 5 yearly	To comply with the HSC Approved Code of Practice, the Control of Asbestos at Work Regulations 1987 and the Control of Substances Hazardous to Health (COSHH) Regulations 1994.
Personnel in cold rooms - safety procedure	1. Inspect and check operation of door release mechanism 2. Check legibility of instruction plate for release mechanism 3. Check door heater 4. Carry out operational check of alarms and emergency lighting	3 monthly	To comply with the Factories Act 1961 and the Electricity at Work Regulations 1989. Compliance with BS 4434 is deemed to satisfy.

Chapter 3   SAFETY PROCEDURES

3.1   Approved code of practice

The Health and Safety Executive has issued an Approved Code of Practice L21, "Management of Health and Safety at Work Regulations 1992". As this is an Approved Code of Practice, non-compliance with any recommendations of the Code can be used as evidence of non-compliance with the Management of Health and Safety at Work Regulations. In all but the most straightforward businesses, it will be necessary to obtain a copy of the Code and implement all the relevant recommendations.

A prime requirement of the Management of Health and Safety at Work Regulations is that every employer shall make a suitable and sufficient assessment of the risks to the health and safety of his employees to which they are exposed while at work and the risks to health and safety to persons not in his employment arising out of, or in connection with, the work of his undertaking. The Code of Practice provides guidance on the assessment of these risks.

A structured approach as below may assist in maintenance management:

Policy statement

As an indication of the senior management or directors' commitment to health and safety at work, it is usual for a policy statement to be drafted and signed by the senior members of the company.

Responsibilities

The particular responsibilities of individuals within the company for safety must be identified, including who has final responsibility for implementing the safety policy of the company.

Determination of risks

Using the guidance given by the Approved Code of Practice the risks to the health and safety of employees and others associated with the work of the company need to be assessed. After carrying out the risk assessment, decisions have to be made about the control measures necessary to manage the risks.

Control measures

Control measures must to be taken to ensure that the identified risks are properly handled. This may require:

1)   Purchase of equipment;
2)   Changes of procedure;
3)   Issuing of instructions;
4)   Training of staff;
5)   Provision of equipment to staff;
6)   Control/feedback facilities.

Safety instructions

Safety instructions, written and formalised, are an effective management tool for the implementation of control measures. They reduce uncertainty about responsibilities, may detail the control measures to be taken and can be used as a reference as well as an instruction to staff on their own responsibilities and the procedures to be followed. They also remind management of their responsibilities to ensure that instructions are complied and that training (and updating) is provided.

Training

The safety instructions will have specified the level of training necessary for staff and this will have been determined before the safety instructions are written up. Employees will normally have been trained as necessary, but the issue of safety instructions will enable this to be checked, both by management and the staff themselves, with further training arranged as required including training in the use of necessary equipment.

Equipment

The safety instructions will make the staff aware of the equipment they should be using and of their responsibility to use equipment provided for them and their responsibilities to keep such equipment in good order, advising supervisors of any deficiencies in the equipment, training and instructions.

3.2 Non-electrical safety instructions

Non-electrical safety instructions are generally outside the scope of this Publication, but the following general safety precautions should be applied :

General principles

Accidents are rarely accidental in the sense of a number of very unlikely circumstances coinciding. They normally arise because basic safety precautions have not been followed, including not using safety equipment provided, e.g. safety glasses or protective footwear. Generally non-electrical instructions for general maintenance work excluding any specific hazard would include requirements for:

1) Protective clothing and equipment;
2) Good housekeeping;
3) Safe access and safe exit; in both normal and emergency conditions;
4) Provision etc. of ladders and scaffolding, or other access equipment;
5) Openings in floors and ceilings;
6) Restricted workspaces;
7) Lighting, both normal and emergency;
8) Lifting and handling;
9) Fire precautions;
10) Hand tools;
11) Mechanical handling devices;

12) Portable power tools;

13) Brazing and welding equipment, including flammable gases e.g. acetylene, dangerous gases e.g. oxygen, and asphyxiating gases e.g. argon.

A typical safety instruction for general and non-electrical maintenance activities is included as Part 2 of Appendix B.

3.3   Electrical safety instructions

Where staff are employed who will be expected to work on electrical installations, they should be issued with safety instructions that have been discussed with and agreed by them. It must be made clear that only staff competent to do so should work on any part of the low voltage electrical installation. Competent staff will be those identified as being competent by the issue to, and formal receipt of, safety instructions. Typical safety instructions are included as Part 3 of Appendix B.

3.4   Live working

Regulation 14 of the Electricity at Work Regulations 1989 states:

> "No person shall be engaged in any work activity on or so near any live **conductor** (other than one suitably covered with insulating material so as to prevent **danger**) that danger may arise unless -
> (a)   it is unreasonable in all the circumstances for it to be dead; and
> (b)   it is reasonable in all the circumstances for him to be at work on or near it while it is live; and
> (c)   suitable precautions (including where necessary the provision of suitable protective equipment) are taken to prevent **injury**."

The Memorandum of Guidance to the Electricity at Work Regulations 1989 advises that work should be carried out live only when there is no other way of reasonably effecting the work. Working includes testing and fault finding. Procedures for working on or near live parts, if allowed at all, must be included in the safety instructions issued to staff. Where the work is of a repetitive nature, such as the routine testing of a particular piece of equipment, it would be reasonable to expect specific test regimes or test equipment to be fabricated that would allow the test and measurements to be carried out without danger to the staff involved.

Chapter 4 ELECTRICAL INSTALLATIONS

4.1 The need for maintenance

Regulation 4(2) of the Electricity at Work Regulations 1989 requires that:

As may be necessary to prevent danger, all systems shall be maintained so as to prevent, so far as is reasonably practicable, such danger.

The Memorandum of Guidance published by the Health and Safety Executive advises that this regulation is concerned with the need for maintenance to ensure the safety of the system, rather than being concerned with the activity of doing the maintenance in a safe manner [which is required by Regulation 4(3)]. The obligation to maintain arises if danger would otherwise result. There is no specific requirement to maintain as such, what is required is that the system be kept in a safe condition. The quality and frequency of the maintenance must be sufficient to prevent danger so far as is reasonably practicable. The HSE Memorandum advises that regular inspection of equipment is an essential part of any preventive maintenance programme. Practical experience of use of the installation may indicate an adjustment to the frequency at which preventive maintenance is to be carried out. This is a matter for the judgement of the duty holder who should seek all the information he needs to make this judgement including advice from equipment manufacturers.

4.2 Fixed installations, equipment and appliances

Regulation 4(2) of the Electricity at Work Regulations makes reference to all systems being maintained so as to prevent, so far as is reasonably practicable, such danger. Systems are defined as follows:

"System" means an electrical system in which all the electrical equipment is, or may be electrically connected to a common source of electrical energy and includes such source and such equipment (Regulation 2(1)).

"Electrical equipment" includes anything used, intended to be used or installed for use, to generate, provide, transmit, transform, rectify, convert, conduct, distribute, control, store, measure or use electrical energy. (Regulation 2(1)).

It is clear that as a consequence of these definitions, "system" includes all electrical equipment from the generating equipment, the fixed wiring of a building, and all the equipment in the building including fixed, portable and hand-held appliances. Electrical equipment includes anything powered by whatever source of electrical energy including battery-powered.

The scope of this Publication can be considered to include:

> distribution systems
> electrical installations (they may not be in buildings);
> electrical equipment supplied from electrical installations.

The maintenance requirements of distribution systems are generally dealt with in Chapters 7 and 8.

There is generally a distinction drawn between the fixed electrical installation of the building covered by BS 7671: 1992 Requirements for Electrical Installations and other items supplied from the fixed installation including appliances. The inspection, test and maintenance requirements of the fixed electrical installation as covered by BS 7671 are discussed later in this Chapter. The in-service inspection and testing of other electrical equipment including appliances, is discussed in Chapter 12 - Equipment and Appliances.

4.3   Frequency of inspection and test of fixed installations

Section 6 of the Health and Safety at Work etc. Act 1974 states that "it shall be the duty of any person who designs, manufactures, imports or supplies any article for use at work to take such steps as are necessary to secure that persons provided with adequate information about the use for which the article is designed or has been tested and about any conditions necessary to ensure that it will be safe and without risks to health". There is a clear duty upon designers and installers of electrical installations to provide information on the need or otherwise for inspection and testing and its frequency. When an installation is designed and installed, assumptions are made by the designer and installer as to what is likely to be the use and abuse of a system. The designer will have assumed certain intervals between inspections, and inspections and tests in his design. The Electrical Installation Certificate of Appendix 6 of BS 7671 requires that the interval at which the installation must be inspected and tested be inserted.

This is clearly a maximum period and experience will show whether the intervals can be extended or need to be shortened. In the absence of guidance from the original designer and installer, minimum periods between inspections are given in Table 4B.

4.4   Inspection and testing regimes

The Management of Health and Safety at Work Regulations requires employers to give consideration as to how they are going to manage health and safety matters. This applies to electrical installations as much as to any other safety matter. Detailed inspection and testing, however thorough and expensive, say every five years is not going to ensure the continuing safety of an electrical installation which might be damaged on a daily basis. Consideration needs to be given to a maintenance regime. General domestic installations have a recommended maximum period between inspections of ten years, but it is presumed that the householder will naturally identify any faults and breakages and arrange to have them repaired in the periods between inspections. The periods between inspection of column 3 of Table 2B or column 3 of Table 4B are obviously too long if defects are not rectified between times. These inspections are to determine if there has been deterioration in the installation and whether changes are necessary to bring the installation into line with the current standard. In the workplace it may not be reasonable to expect routine reporting of defects so regular routine checks must be carried out. These are for breakage and wear. Breakages and deterioration due to wear and tear cannot be left for such periods. In areas open to the public where defects might not be reported, further checks see column 2 of Table 4B must supplement the inspections of column 3.

Table 4A   Routine Checks

Activity	Check
Defects Reports	All reported defects have been rectified

Inspection	Look for:
	    breakages
	    wear/deterioration
	    signs of overheating
	    missing parts (covers, screws)
	    switchgear accessible (not
	    obstructed)
	    doors of enclosures secure
	    adequate labelling
	    loose fixings

Operation	Operate:
	    switchgear (where reasonable)
	    equipment - switch off and on
	    including RCDs (using test button)

The frequency of routine checks will depend upon the circumstances and they do not necessarily need to be carried out by electrically skilled staff. Frequent, even daily checks may be appropriate particularly in areas open to the public. Table 4A summarises the activities of a routine check and the defects looked for.

Table 4B: Frequencies of Inspection of Electrical Installations

Type of installation 1	Routine check Section 4.4 2	Maximum period between inspections and testing as necessary 3	Reference (see notes below) 4
General installation			
Domestic	----	Change of tenancy/10 years,	1, 2
Commercial	1 year	Change of tenancy/ 5 years,	1, 2
Educational establishments	4 months	5 years	
Hospitals			
Industrial	1 year	5 years	1, 2
Residential accommodation	1 year	3 years	1, 2
Offices	at change of occupancy/1 year	5 years	1
Shops			
Laboratories	1 year	5 years	1, 2
	1 year	5 years	1, 2
	1 year	5 years	1, 2
Buildings open to the public			
Cinemas	4 months	1 year	2, 6, 7
Church installations	1 year	5 years (quinquenially)	2
Leisure complexes	4 months	1 year	1,2,6
Places of public entertainment	4 months	1 year	1, 2, 6
Restaurants and hotels	1 year	5 years	1,2
Theatres	4 months	1 year	2, 6, 7
Public houses	1 year	5 years	1,2, 6
Village halls/Community centres	1 year	5 years	1, 2
External installations			
Agricultural and horticultural	1 year	3 years	1, 2
Caravans			
Caravan Parks	1 year	3 years	
Highway power supplies	6 months	1 year	1, 2, 6
Marinas	as convenient	6 years	
Fish farms	4 months	1 year	1, 2
	4 months	1 year	1, 2
Special installations			
Emergency lighting	Daily/monthly	3 years	2, 3, 4
Fire alarms	Daily/weekly/monthly	1 year	2, 4,5
Launderettes	1 year	1 year	1, 2, 6
Petrol filling stations	1 year	1 year	1, 2, 6
Construction site installations	3 months	3 months	1, 2

Reference Key

1. Particular attention must be taken to comply with S1 1988 No 1057. The Electricity Supply Regulations 1988 (as amended).
2. S1 1989 No 635. The Electricity at Work Regulations 1989 (Regulation 4 and Memorandum).
3. See BS 5266: Part 1: 1988 Code of practice for the emergency lighting of premises other than cinemas and certain other specified premises used for entertainment.
4. Other intervals are recommended for testing operation of batteries and generators.
5. See BS 5839: Part 1: 1988 Code of practice for system design installation and servicing (Fire detection and alarm systems for buildings).
6. Local Authority Conditions of Licence.
7. S1 1995 No 1129 (Clause 27) The Cinematography (Safety) Regulations.

4.5   Periodic inspection and testing

The requirements for periodic inspection and testing are provided in Chapter 73 of BS 7671. It is worthwhile considering exactly what the requirements are:

Regulation **731-01-02** states:

"Inspection comprising careful scrutiny of the installation shall be carried out without dismantling or with partial dismantling as required, supplemented by testing to verify compliance with Sections 731 and 732 and as far as possible to provide for:

(i)   the safety of persons and livestock against the effects of electric shock and burns, in accordance with Regulation 120-01, and

(ii)   protection against damage to property by fire and heat arising from an installation defect, and

(iii)   the identification that the installation is not damaged or deteriorated so as to impair safety, and

(iv)   the identification of installation defects or non-compliance with the requirements of the Regulations which may give rise to danger."

It should be noted that the requirement is for careful scrutiny (inspection) supplemented by testing as necessary, not for inspection and testing. The intent is that, where possible, the installation should not be dismantled as this obviously introduces the risk of its not being correctly reassembled and that any partial dismantling be carried out only as required. The careful scrutiny is to be "supplemented" by testing to verify compliance with quite general requirements, to provide for the safety of persons and livestock against electric shock and burns, and the protection of property from the risk of fire.

For an installation that by inspection alone would appear to be in good condition, little further is to be gained by carrying out insulation resistance measurements of the complete installation. However, if inspection has indicated that the insulation readings are likely to be low, such as might be found with say rubber insulation, it might be reasonable to progress and carry out some insulation resistance measurements. If possible, the general condition of an installation should be assessed by the electrical contractor before he starts the detailed work of inspection and testing. Agreement can then be reached prior to the work proper being commenced as to the testing likely to be necessary. This is obviously preferable to differences arising after completion of the work as to what was contracted to be carried out.

Often, inspection and test work is tendered for competitively, when there needs to be some good guidelines as to what proportion of the installation is to be tested. To assist in these situations, a typical requirement is listed in Table 4C. The guidance in Table 4C applies to installations where no alterations are known to have been made since the last inspection and test. Suitable test methods are described in Chapter 7.

When a periodic inspection and test is carried out the Periodic Inspection Report of Appendix 6 of BS 7671 should be completed by the person carrying out the activity, together with an installation schedule including test results as found in IEE Guidance Note No. 3.

4.6   Minor Works

For minor electrical work that does not include the provision of a new circuit the Minor Electrical Installation Works Certificate of Appendix 6 of BS 7671 is appropriate.

4.7   Certificates

The Certificates of BS 7671 are included in an appendix to the Standard. This means that the forms used do not have to be identical to those in BS 7671. It may be appropriate to use a different layout. An electrician carrying out many minor repairs in a day may well use a form that allows a number of minor electrical works to be recorded on one form (certificate).

Table 4C: Summary of Periodic Testing

Test Type	Recommendation
Continuity	Tests to be carried out between: all main bonding connections all supplementary bonding connections **Note:** When an electrical installation cannot be isolated protective conductors, including bonding conductors, must not be disconnected.
Polarity	Tests to be carried out at: the origin of the installation all socket-outlets 10% of control devices including switches 10% of centre-contact lampholders **Note:** If failures found, a full test is made on that particular part of the installation and testing on the remainder increased to 25%. If a further fault is found the complete installation is tested.

Earth fault loop impedance	Tests to be carried out:
	at the origin of the installation;
	at each distribution board and either;
	    at each socket-outlet;
	    at the extremity of every radial circuit.
	    (note 1)

Insulation resistance	If measurements are to be made then test : the whole installation with all protective devices in place and all switches closed. Where circuits include electronic devices, only a test between phase and neutral conductors connected together and protective conductors shall be made. (See 4.5)
Functional	Activities to be carried out :- all devices for isolation and switching to be operated all labels to be checked all interlocking mechanisms verified the operation of all RCDs to be verified by the simulation of a fault (note 2) or by a test instrument, and by the operation of the test button all manually operated circuit-breakers to be switched to verify the devices open and close all circuit-breakers with injection testing facilities are to be injection tested to confirm settings and operation.

Notes

1)   Most earth loop impedance testers will trip RCDs in the circuit. If this occurs loop impedance measurement can be made by the measurement of $R_1 + R_2$ during continuity testing added to the measured value of Z_e or by shorting out the RCD.

2)   Simulation of a fault is the requirement of BS 7671 Regulation 713-12-02. A fault can be simulated with a test lamp, a 15 W pygmy lamp providing 60 mA to earth.

Chapter 5   EMERGENCY LIGHTING

5.1   General

Emergency lighting is provided to prevent a hazard in the event of the loss of supply to the normal lighting installation. Whilst in general, emergency lighting is considered to be escape lighting, when carrying out maintenance inspections of emergency lighting installations all hazards that might arise as a result of loss of the normal lighting must be considered. Emergency lighting may be required to illuminate switchrooms and control rooms, to facilitate restoration of supplies or management of facilities to allow dangerous plant to be shut down. One of the most important functions of the emergency lighting is to provide reassurance to occupants and to allow orderly and speedy evacuation of a building, should this be necessary.

Emergency lighting for areas where artificial lighting is required night and day has to be considered from basic principles. There may be a need to illuminate the area to prevent danger in the event of loss of electricity supply as well as a need to provide escape lighting.

Escape lighting must:-

1)   Indicate the escape routes;
2)   Illuminate the escape routes;
3)   Illuminate fire alarm call points and fire fighting equipment.

The requirements for escape lighting are detailed in BS 5266: Part 1 : 1988. This Standard is essential for anyone designing an escape lighting system. The guidance given in this chapter is based upon that given in BS 5266.

Note: British Standard Code of Practice CP 1007: 1995, Maintained lighting for cinemas relates to the safety lighting and management lighting in cinemas.

5.2   Siting of escape luminaires

It is necessary to install an escape lighting luminaire at each exit door, particularly emergency exit doors and at any other location that will aid escape, facilitate initiation of alarm, and identification of fire equipment. Such locations include:

1)   Intersection of corridors;
2)   Exit doors;
3)   Change of direction of escape route (other than staircases);
4)   Staircases;
5)   The changes of floor level;
6)   Outside final exits;
7)   Near each fire alarm call point;
8)   Near fire fighting equipment;
9)   Safety signs, including exit signs.

The locations listed above are an essential requirement of BS 5266. Illumination is recommended additionally in the following locations:

1) External areas in the immediate vicinity of exits;
2) Lift cars;
3) Moving stairways and walkways;
4) Toilets, lobbies and closets exceeding 8m² where there is no borrowed light from other emergency lighting via windows, doorways, etc.

5.3 Fire safety signs

The Health and Safety (Safety Signs and Signals) Regulations 1996 implement European Council directive 1992/58EEC specifying minimum requirements for the provision of safety signs at work. This legislation is general, having requirements for all safety signs - prohibitory, warning and mandatory. It includes requirements in part 3 for fire safety signs. The emergency escape and first aid signs described in the Health and Safety Executive's guidance on the regulations are shown in Figures 5.1 and 5.2 and the fire fighting signs in Figure 5.3.

Sign colours

The regulations specify the colour of signs as follows:

Colour	Meaning or purpose	Instruction or information
Red	Prohibition sign	Dangerous behaviour
Red	Fire fighting equipment	Identification of location
Red	Danger alarm	Stop shutdown emergency cut-out devices, evacuate
Green	Emergency escape, first aid sign	Doors, exit routes, equipment facilities

Emergency signs and emergency lighting should generally be positioned between 2 and 2.5m above floor level. The effect of smoke must be considered and this will mitigate against signs or luminaires at a higher level. All exit, emergency exit and escape route signs need to be illuminated so that they are legible at all times. In the event of failure of the normal electricity supply, such signs should remain illuminated. The illumination may be from:

a) An external luminaire. Legends should comply with the Health and Safety Executive recommendations reproduced in Figure 5.1;

b) An internally illuminated sign, generally constructed in accordance with BS 5499 Part 1.

Legends in accordance with BS 5499 will meet the requirements of the new regulations provided they continue to fulfil their purpose effectively. However, the markings should generally be as Figure 5.1;

c)   Self-luminous e.g. tritium tube, signs in accordance with BS 5499 Part 2, but again with legend as recommended in the Health and Safety Executive guidance reproduced in Figure 5.1.

Figure 5.1   Emergency/escape route signs

Intrinsic features:

(a)   rectangular or square shape
(b)   white pictogram on a green background (the green part to take up at least 50% of the area of the sign)

Emergency exit/escape route signs

Chapter 11 provides information on the exit signs adopted generally in the UK.

Figure 5.2 Further directional and first-aid signs

Supplementary 'This way' signs for emergency exits/escape routes

First-aid signs

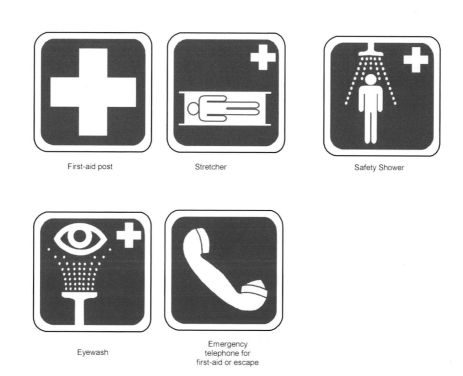

Figure 5.3   Fire fighting signs

Intrinsic features:

(a)   rectangular or square shape

(b)   white pictogram on a red background (the red part to take up at least 50% of the area of the sign)

Fire fighting signs

Fire hose   Ladder

Emergency fire
telephone   Fire extinguisher

Supplementary 'This way' signs for fire fighting equipment.

5.4   Installation requirements

Escape route luminaires may be either self-contained, that is take their supply from the normal electricity supply when it is available but using internal batteries on the failure of the supply, or supplied from a central battery system. Luminaires should be constructed in accordance with relevant British Standards and be suitable for the environment.

Wiring

It is important to note that wiring to self-contained luminaires is not considered to be part of the emergency lighting circuit. This means that such wiring does not have to comply with the requirements for cables of BS 5266.

However, wiring connecting escape lighting (sign or escape route illumination) to the emergency supply should possess inherently high resistance to both fire and physical damage or be suitably protected. The basic requirement is that cables should be routed through areas of low fire risk. In general the choice is between :

1)   Mineral-insulated cable to BS 6207;
2)   Cables complying with category B of BS 6387 : Specification for performance requirements for cables required to maintain circuit integrity under fire conditions.
       Cables to the following standards meet this requirement :
       BS 7629 : Specification for thermosetting insulated cables with limited circuit integrity when affected by fire.
       BS 7846 : 600/1000V Armoured cables having limited circuit integrity and low emission of smoke and corrosive gases when affected by fire.

BS 5266 states that other wiring systems including PVC-insulated and sheathed cables in rigid PVC conduits, PVC-insulated cables in steel conduit, and PVC-insulated and sheathed steel-wire armoured cables, require additional fire protection. The additional fire protection could be provided by burying the cables in the structure of the building or installing the cables where there is a negligible fire risk and separated from any significant risk by a wall, partition or floor complying with any of the following:

(i)   specifications tested or assessed under the appropriate Part of BS 476;

(ii)   appropriate British Standard specifications or codes of practice;

(iii)   specifications referred to by building regulations applicable to the building;

(iv)   cables enclosed in steel conduit complying with the tests given in BS 6387 for fire resistance.

Where appropriate, requirements for stability, integrity and insulation must be met. The test by fire is applied to the side of the construction remote from the cable. In certain premises a longer duration of fire resistance may be necessary for escape purposes.

System supplies

The normal supply should be so arranged that continuity of supply is assured in all but the most extreme circumstances. If, for example, it is the practice to switch off the supply to a premises when not in use or at night, the design of the installation should be such that the supply to the emergency lighting is maintained. This is essential to ensure that emergency lighting batteries remain charged, and are not run down by their supply being disconnected for long periods. This is, of course, as necessary for self-contained luminaires as for a central battery.

Isolator switches

To reduce the likelihood of the inadvertent disconnection of the supply to the emergency lighting system, isolator switches and protective devices should be installed in a location inaccessible to unauthorised persons, or should be protected against unauthorised operation. Inadvertent disconnection might result in the rundown of the emergency batteries or their premature failure. Each isolator switch, protective device, key and operating device should be marked:

Emergency/Escape/Standby lighting/Do not switch off

as appropriate.

It is important to give careful consideration to the protective measures selected for emergency lighting supplies. Continuity of supply is particularly important, so precautions must be taken against unwanted operation of a circuit protective device, for example, at switch-on due to surges or as a result of earth leakage.

Supplies for safety services

Chapter 56 of BS 7671: 1992 Requirements for Electrical Installations specifies requirements for supplies for safety services.

5.5   Luminaire illumination duration

BS 5266: Part 1 recommends categories of emergency lighting luminaire for various premises. These are summarised in Table 5A.

Two basic types of luminaire are specified - maintained and non-maintained.

Maintained emergency lighting (M) - a lighting system in which all emergency lighting lamps are in operation at all material times.

Non-maintained emergency lighting (NM) - a lighting system in which all emergency lighting lamps are in operation only when the supply to the normal lighting fails.

Self-contained emergency luminaires may be either maintained or non-maintained. They are luminaires which contain all the elements of the emergency luminaire, including the battery, the lamp, the control unit, test and monitoring facilities. BS 5266 categorises luminaires by there being maintained or non-maintained and by the number of hours which they can maintain their light output to an acceptable level after the failure of supply.

For example, a non-maintained luminaire with a duration of two hours is given the designation NM/2 (see note to Table 5A).

In general, the Standard advises that it is unlikely that evacuation will take longer than one hour and this generally should be sufficient time for the illumination. However, in certain conditions the local authority licence may allow a period of time for continued occupation after the failure of the normal lighting. In these situations, the minimum duration of the emergency lighting should be one hour after any such period of permitted occupation. Additionally, particularly in larger premises, emergency lighting will be necessary after the evacuation of the building has been substantially completed, for safety requirements such as searching of the premises to ensure that no persons have been left behind or to allow reoccupation of the premises in order to get people off the street and into a place of relative safety.

Table 5A   Emergency lighting duration category

TYPE OF PREMISES			MINIMUM CATEGORY Note 1
1. Residential	With sleeping accommodation:	10 bedrooms or more	minimum of NM/3 or M/3
	e.g. hospitals, nursing homes, hotels, guest houses, colleges with boarders, boarding schools	Small - not more than 10 bedrooms or not more than one floor above or below ground floor level	minimum of NM/2 or M/2
2. Non-residential	Treatment or care : e.g. special schools, clinics		minimum of NM/1
3a. Non-residential	Recreational: theatres, concert halls	Premises where there is facility to dim the normal lighting or for the consumption of alcohol	M/2
	exhibition halls		
	sports halls	Other	M/2 or NM/2
	public houses restaurants	Small - no more than 250 persons present	M/1 or NM/1
3b. Non-residential	Recreational: Ballrooms and dance halls Cinemas (licensed under the Cinemas Act 1985) Bingo premises (licensed under the Gaming Act 1968 as amended) Ten pin bowling premises		As 3a above but also comply with British Standard Code of Practice CP 1007:1995 Maintained lighting for cinemas
4. Non-residential	schools, colleges laboratories		M/1 or NM/1
5. Non-residential	Public premises: e.g. town halls libraries offices	Premises where lighting may be dimmed	M/1
	shops art galleries museums	Other	M/1 or NM/1
6. Non-residential	Industrial: e.g. manufacturers provisions storage		Minimum of NM/1 or M/1

Note

M signifies maintained emergency lighting - the emergency lighting lamps are in operation at all material times.

NM signifies non-maintained emergency lighting - the emergency lighting lamps are in operation only when the supply of the normal lighting fails. Self-contained luminaires may be maintained or non-maintained.

The digit indicates period of illumination after failure of supply e.g. NM/2 is a non-maintained luminaire with a duration of 2 hours.

5.6   Maintenance

After installation of an emergency lighting system, a completion certificate should have been supplied to the person requesting the work. Maintenance of an emergency lighting system should be carried out on daily, monthly, six-monthly and three-yearly basis, as described below. The results of these maintenance activities should be recorded in a log book and kept available for examination. A log book should record :

a)   Date of any completion certificate, including certificates relating to extensions or alterations;
b)   The date of each periodic inspection and test certificate;
c)   Date and brief details of each service, inspection or test;
d)   Dates and brief details of all defects and the remedial action taken;
e)   Date and brief details of all alterations to the emergency lighting installation.

Daily activities should include :

a)   Faults noted in the log book have been actioned;
b)   Lamps in maintained luminaires are lit;
c)   The main control or indicating panel of central battery systems or engine-driven plant indicates a healthy state;
d)   All faults found are recorded in the log book.

Monthly

A typical monthly periodic inspection and test certificate is shown as Figure 5.4. Monthly maintenance should include:

a)   the energising of all self-contained and illuminated exit signs by simulation of a failure of the supply for a sufficient period only to ensure that each lamp is illuminated.

It is important that this period should be as short as possible to avoid discharging the batteries - it should under no circumstances exceed one quarter of the rated duration of the luminaire or sign. At the end of the test the supply to the normal lighting should be restored and a check made that any indicator lamps indicate a healthy state.

b)   Each central battery system energised from its central battery by simulation of a failure. Again, the period of simulated failure should be minimised and should not exceed a quarter of the rated duration of the battery capacity.

If it is not possible to examine all the luminaires and/or signs in a quarter of the rated duration time, the test must be repeated after the battery has been fully recharged. Again, at the end of each test the supply is returned to normal.

c)   Engine-driven generating plants should be started by simulation of mains failure and allowed to energise emergency lighting for a period of at least one hour.

d)   For engine-driven generating plant with back-up batteries, the system should be tested to check the battery supply, by simulation of failure of the supply with the engine-driven generating plant prevented from starting. After this check, the starting of the engine should be allowed in the normal manner, and the emergency lighting system run from the generating plant for at least one hour. At the end of the test period the emergency equipment should be returned to normal and healthy indications checked. Fuel tanks should be filled, oil and coolant levels topped and other maintenance activities associated with the prime mover carried out as necessary.

Six-monthly

At six-monthly intervals, the monthly test should be carried out plus:

a)   Each three-hour self-contained luminaire and internal illuminated sign should be energised from its battery for one hour by simulation of a fault. If the luminaire is rated as having a duration of one hour, then the period of simulated failure should be 15 minutes.

b)   Each three-hour central battery system should be continuously energised from its battery for one hour by simulation of a fault. If the system is rated as having a duration of one hour then the period of simulated failure should be 15 minutes.

c)   Engine-driven plant with back-up batteries should be tested allowing supply from the back-up batteries for a period of one hour. On restoration of the normal supply checks should be made that healthy indications are given for all alarms. Fuel tanks should be filled, oil and coolant levels etc. topped up as necessary.

Three-yearly

The monthly inspection should be carried out with the following additional tests:

a)   The basic design of the installation should be confirmed as outlined with the notes to the model form.

b)   Each self-contained luminaire or internally illuminated sign should be tested for its full duration.

c)   Each central battery system should be tested for its full duration.

d)   Each generator back-up battery should be tested for its full duration.

At the end of these tests the system should be returned to normal, healthy signals checked and fuel tanks, oil coolant levels topped up etc. as necessary.

5.7 Self-contained luminaires with sealed batteries

It is recommended that self-contained luminaires with sealed batteries over 3 years old should be given the three-yearly test each year. This is necessary to check that the sealed batteries have an adequate storage capacity.

Figure 5.4 Emergency lighting - Periodic inspection and test certificate

Occupier/owner...

Address of premises...

.... Tel no...........................

Date of inspection and test..

Inspection and test carried out by ....

Name and address ....

.... Tel no...........................

I/We hereby certify that the emergency lighting installation at the above premises has been inspected and tested in accordance with the schedule below by me/us and to the best of my/our knowledge and belief complies at the time of my/our test with the recommendations of BS 5266 'Emergency lighting' Part 1 : 1988 'Code of practice for the emergency lighting of premises other than cinemas and certain other specified premises used for entertainment', published by BSI, for a category...* installation, except as stated below.

Signature of person responsible for inspection and test ....

Qualification† ....

Date...

For and on behalf of...

Details of variations from the code of practice (BS 5266: Part 1: 1988):

Schedule to emergency lighting periodic inspection and test certificates

NOTE 1. Because of the possibility of failure of the supply to the normal lighting occurring shortly after a period of testing, all tests should be undertaken at times of minimum risk. Alternatively, suitable temporary arrangements should be made until the batteries have been recharged.

NOTE 2. The figures in brackets indicate the relevant clauses of BS 5266: Part 1: 1988.

Results of inspection and tests

(a) Are correct entries made in the log book?	YES/NO
(b) Are record drawings available?	YES/NO

*Enter M/1, 2 or 3 or NM/1, 2 or 3 as appropriate (see **6.12** of BS 5266: Part 1: 1988).

†Qualifications: a suitably qualified electrical engineer or a member of the Electrical Contractors' Association or the Electrical Contractors' Association of Scotland; or a certificate holder of the National Inspection Council for Electrical Installation Contracting; or a qualified person acting on behalf of one of these (in which case it should be stated on whose behalf he is acting). Where acceptable to the enforcing authority the authorised representative of a manufacturer of emergency lighting equipment may be deemed to be a suitably qualified person.

(from BS 5266)

Model periodic inspection and test certificate *(concluded)*

(c)	Are record drawings correct?	YES/NO
(d)	*Signs.*	
	(1) Are the signs correctly positioned? (**6.9**)	YES/NO
	(2) Are details of the signs correct? (**6.9**)	YES/NO
	(3) Do the self-luminous signs (if any) need changing before the date of the next scheduled inspection? If so state date...................................(See label on sign) (**6.9**)	YES/NO
(e)	*Luminaires.* Are luminaires correctly positioned? (**6.7, 6.8** and **10.3**)	YES/NO
(f)	*Illumination for safe movement* (clause **5** and see record drawings).	
	(1) Are the correct lamps installed in the luminaires? (**6.13**)	YES/NO
	(2) Has there been any change in the decor or layout of the premises since the last inspection, which has caused any significant reduction in the effectiveness of the lighting system?	
	(Any changes to be stated under COMMENT below.)	YES/NO
	(3) Is the installation in a generally satisfactory condition?	YES/NO
(g)	*Marking.*	
	(1) Are the category and nominal operating voltage of the system clearly marked or readily identifiable? (**6.13**)	YES/NO
	(2) Are luminaires clearly marked to indicate the correct lamp for use? (**6.13**)	YES/NO
	(3) Is information available to ensure correct battery replacement? (**6.13**)	YES/NO
(h)	*Wiring systems* (clause **8**).	
	(1) Are the results recorded on the last inspection and test certificate satisfactory?	YES/NO
	(2) State the date of this inspection and test	
(i)	*Power services.*	
	(1) Are the charging arrangements for batteries satisfactory? (**6.11, 12.2** and **12.4**)	YES/NO
	(2) Do changeover devices operate satisfactorily upon simulation of failure of the normal supply?(**6.11** and **12.4**)	YES/NO
(j)	*Central battery systems including backup batteries.*	
	(1) After operation for the rated duration:	
	(i) do all luminaires operate? (**6.7, 6.8** and **12.4**);	YES/NO
	(ii) are all signs illuminated and visible? (**6.9** and **12.4**);	YES/NO
	(2) Following the restoration of the system to normal:	
	(i) is the battery charger functioning? (**6.11** and **12.4**);	YES/NO
	(ii) are the levels and the specific gravities of the battery electrolytes satisfactory, where applicable?	YES/NO
(k)	*Engine-driven generating plant.*	
	(1) After a period of operation of at least 1 h:	
	(i) do all luminaires operate? (**6.7, 6.8** and **12.4**);	YES/NO
	(ii) are all signs illuminated and visible? (**6.11** and **12.4**);	YES/NO
	(iii) does the back-up battery, where installed, operate satisfactorily? (see (j) above)	YES/NO
	(2) Following the restoration of the system to normal:	
	(i) is the battery charger for the engine starter functioning? (**6.11** and **12.4**)	YES/NO
	(ii) are the fuel, coolant and lubricating oil levels correct? (**12.4**)	YES/NO
(l)	*Self-contained luminaires and signs.*	
	(1) After operation for the rated duration, does each self-contained luminaire and sign operate? (**6.9, 6.11** and **12.4**)	YES/NO
	(2) Following restoration of the system to normal supply, is the battery charger functioning? (**6.11** and **12.4**)	

COMMENT (if any) and variation from the code of practice.

(from BS 5266)

Chapter 6   FIRE DETECTION AND ALARM SYSTEMS

6.1   Standards

Those persons with responsibilities for design, installation or maintenance of fire detection and alarm systems in buildings should obtain BS 5839: 1988, Fire detection and alarm systems for buildings, Part 1: Code of practice for system design, installation and servicing. The Guidance given in this Chapter is based upon this standard. (Guidance for buildings comprising a number of self-contained dwelling units is given in Part 6 of BS 5839).

6.2   Types of system

Fire detection and alarm systems are categorised as shown in Table 6A.

Table 6A   Fire detection and alarm system categories

Type	Description
P	automatic detection and alarm systems for the protection of property only
P1	detection and alarm systems installed throughout the protected building
P2	detection and alarm systems installed only in defined parts of the building
L	automatic detection and alarm for the protection of life
L1	detection and alarm systems installed throughout the protected building
L2	detection and alarm systems installed only in defined parts of the protected building Note: A type L2 system should include the coverage required for type L3 system.
L3	detection and alarm systems installed only for the protection of escape routes
M	manual alarm systems. (These are not further sub-divided.)

Note. In BS 5839 Part 6 an additional letter 'D' is added to indicate systems for Dwellings e.g. L2 is designated LD2.

When deciding upon the type of system to be installed, the escape procedures to be followed after the alarm has been initiated must be known. The design will have to take into account the mode of evacuation of the building, including the need for advice to be given to those in occupation. This may require a public address system. In a large building where evacuation must be controlled, it may very well be necessary to segregate the alarm and voice systems to provide for orderly evacuation.

Note should be made of Type L3 systems where detection and alarms are installed only for the protection of escape routes. Warning of fire is given only in time for the escape routes to be used before they become blocked by heat or smoke. A Type L3 system should not be expected to protect people who might be involved with the fire at ignition or in its early stages.

6.3   Power supplies

Fire detection and alarm systems should be connected to the mains supply by their own isolating device. The cover of the isolating device should be coloured red and labelled as follows:

> FIRE ALARM: DO NOT SWITCH OFF

The isolating device should be secured from unauthorised operation to prevent unwanted switching. The British Standard advises that the device may be contained in a securely closed container with a breakable cover.

If the isolating device is supplied from the live side of a supply so that it is not isolated by the incoming main switch a further label should be fixed as follows:

> Warning: This supply remains live when the main switch is turned off

There should be a further label fixed to the main switch as follows:

> Warning: The fire alarm supply remains live when this switch is turned off

If the fire detection and alarm system is supplied by the main switch this switch should be labelled as follows:

> Warning: This switch also controls the supply to the fire alarm system

It is a general rule that any switch that can disconnect the power supply to all or part of a fire detection and alarm system should be coloured red and labelled:

> FIRE ALARM: DO NOT SWITCH OFF

6.4 Cable types

The operation of a fire detection and alarm system is very dependent upon reliable connections between the various components so that supplies and signals are maintained and transmitted. Maintenance is carried out to ensure, as far as is reasonably possible, that all parts of the system including the cabling are in a working condition should a fire occur.

The design of the system should be such that after the initiation of an alarm, even if the connections and supplies to detectors and alarm initiation points are destroyed, the fire alarms should continue to sound. This means that power supplies to control equipment and to the alarm sounders should allow prolonged operation during the fire. Generally, it is required that cables and connections between sounders, control and indicating equipment, and power supplies should resist a fire for at least ½ hour. Cable types recommended to provide this prolonged operation include:

a)   mineral insulated copper sheathed cables to BS 6207

b)   cables complying with BS 6387 meeting category requirements AWX or SWX.

Other cables can provide prolonged operation if they are protected against exposure to fire by either:

a)   burial in the structure of the building and protection by the equivalent of at least 12 mm of plaster or

b)   separation from any significant fire risk by a fire wall, partition or floor having at least ½ hour fire resistance.

As a general requirement cables should be routed through areas of low fire risk.

6.5 Segregation of cables

As well as having good fire performance it is recommended that the fire alarm cables be segregated from those of all other systems to prevent electrical faults damaging the fire detection and alarm cables. In particular, fire alarm power and signal cables should be separated from other systems by one or more of the following methods:

1)   Installation in conduit, ducting or trunking, or a channel reserved for fire alarms.

2)   Mechanically strong rigid continuous partitions of a non-combustible material.

3)   Spacing at least 300 mm from conductors and cables of other systems.

4)   Wiring in cables complying with BS 6387 incorporating an earthed metallic sheath and having an overall insulating sheath e.g. cables complying with BS 7629.

5)   Wiring in mineral insulated copper sheathed cables with an insulating sheath or barrier.

6.6   User responsibilities for supervision

The owner or other person responsible for the premises should appoint a responsible person to superintend the system. The person should have the authority to ensure that:

1. The fire detection and alarm system is maintained in working order by the adoption of the maintenance and servicing procedures.

2. Escape procedures are implemented, including practice procedures, so that people can be safely evacuated in the event of an alarm. These procedures should be agreed with the fire authority.

3. All staff are instructed and practised in proper actions to be taken in the event of a fire and staff with particular responsibilities are also suitably trained.

4. Persons who might by the nature of their work interfere with or degrade the fire alarm, or fire protective systems by changes of any kind, should be required to obtain the consent of the person responsible for the fire detection and alarm system.

5. All passageways, paths or other avenues of escape, and access to fire alarm and fire fighting equipment, are kept clear at all times.

6.7   Records

Up-to-date record drawings and operating instructions should be maintained and preferably retained in the same location as the control and indication equipment.

The responsible person should ensure that a log book is kept which would include the following:

a)   Name, location, extension number of the responsible person
b)   The servicing agent or servicing arrangements
c)   Record of all alarms (false, test, practice and genuine), including the location of the device initiating an alarm
d)   Dates, times and types of all defects and faults
e)   Date and time when defects and faults rectified
f)   Dates and types of all tests
g)   Dates and types of all services
h)   Dates and times of all disconnection and disablements
i)   All alterations to the system.
j)   A record of the names of persons/organisations carrying out tests, repairs, alterations or services of any kind.

6.8   Prevention of false alarms

The responsible person should ensure and have the responsibility and authority to ensure that all staff, contractors and others visiting the premises are aware of the alarm system and the precautions they must take to avoid false alarms.

Notices should be displayed at all entrance areas where detectors are sited to advise of the installation.

All tenders for contracts, and orders for work, should provide advice and precautions to prevent accidental damage to or maloperation of the alarm system. Where building works etc. are undertaken, precautions shall be taken to prevent damage to the alarm system, prevent maloperation and limit the zones disconnected.

6.9   Servicing

The prevention of false alarms

It is important that testing and maintenance of fire alarm systems does not result in false alarms. It is particularly important that automatic transmission of 999 dialling is prevented before a routine test.

Daily servicing

Daily checks of the following should be made:

a)   The alarm panel indicates normal operation or if not any fault has been recorded
b)   Any fault identified the previous day has been rectified
c)   Connections to the public fire brigade or remotely manned centre are tested unless there is continuous monitoring.

Weekly servicing

Each week the following servicing should be carried out:

a)   At least one detector, call point or end of line switch on one zone should, be operated, to test the ability of the control and indicating equipment to receive the signal and sound the alarm.

b)   For systems having thirteen zones or less, each zone should be tested in turn. If there are more than thirteen zones, then more than one zone must be tested in any week so that the interval between tests does not exceed 13 weeks.

c)   If the batteries are accessible then a visual examination of the battery and its connections should be made. Any defects should be reported and rectified.

d)   The fuel and lubricating oils and coolant levels of prime movers should be checked and corrected as necessary.

e)   Any other alarm or indicating system requiring operation should be checked.

Any defects should be reported in the log book.

Monthly servicing

Each month the following servicing is recommended:

a)   Automatically started standby generators should be started by simulation of failure of the normal power supply and allowed to run for at least one hour. The fire alarm system should be checked to see if any faults are caused by the use of the generator.

b)   At the end of testing the system must be returned to normal and fuel, oil, and coolant levels checked and topped up as necessary.

Three-Monthly Inspection

The responsible person should ensure that every three months the following checks are carried out:

a)   All faults identified in the log book have been actioned.
b)   The batteries and their connections are examined and tested as specified.

        (i)   Secondary batteries are examined and checked including specific gravities as necessary
        (ii)   Primary batteries should be tested by taking such measurements to ensure that each cell's condition is satisfactory i.e. voltage testing with load applied.

c)   The alarms of the control and indicating equipment function correctly on operation of a detector or call point in each zone.
d)   The operation of the alarm sounders linked to a remote control centre other than 999 auto dialler is tested, after appropriate consultation and agreement.
e)   All ancillary functions of the control panel should be tested.
f)   All fault indicators and their circuits checked, preferably by simulation of appropriate faults.

A visual inspection of the protected building should be carried out to ascertain if any changes in structure or use have occurred that would require resiting of call points, detectors or sounders.

Defects should be recorded in the log book and reported to the responsible person and action taken to correct.

Annual inspection and test

The quarterly inspections and tests should be carried out each year, plus:

a)   Each detector should be checked for correct operation
b)   Visual inspection should be made of all cable and fittings to check that they are secure
c)   All equipment should be checked for signs of damage, ageing or any other indication of the likelihood of malfunction.

All the defects should be recorded in the log book and action taken to correct recorded.

Wiring maintenance

The hard wiring of a fire detection and alarm system should be checked at least as frequently as the fixed electrical installation and at intervals not exceeding five years, in accordance with the recommendations for periodic inspection and testing of BS 7671.

On completion of the work, a certificate of test should be presented to the responsible person by the person carrying out the work.

Certification

New detection and alarm systems on completion of inspection and testing should be certified as per Figure 6.1. After each periodic inspection and test of the system a proforma as Figure 6.2 should be completed.

Model log book

A typical log book recording system is shown in Figure 6.3.

Figure 6.1   Model certificate of installation and commissioning of a fire alarm system

---

Certificate of installation and commissioning of the fire alarm system at :

Protected area ....

Name of occupier ....

Address ....

....

My attention has been drawn to the recommendations of BS 5839 : Part 1 : 1988; in particular, to clauses 14 (false alarms), 28 and 29 (user responsibilities).

In accordance with BS 5839: Part 1: 1988, subclause 26.1, record drawings and operating instructions have been supplied and received by:

Signed   ....Status .... Date.....................

For and on behalf of (user) ....

....

In accordance with BS 5839: Part 1: 1988, subclause 26.2, the installation has been inspected and been found to comply with the recommendations of the code.
In accordance with BS 5839: Part 1: 1988, subclause 26.3, the insulation of cables and wires has been tested.
In accordance with BS 5839: Part 1: 1988, subclause 26.4, the earthing has been tested.
In accordance with BS 5839: Part 1: 1988, subclause 26.5, the entire system has been tested for satisfactory operation.
In accordance with BS 5839: Part 1: 1988, subclause 26.6, it is certified that the installation complies with the recommendations of the code, other than the following deviations:

Signed (Commissioning engineer) .... Date ....

Name (in block letters) ....

For and on behalf of (installer) ....

The system log book is situated ....
The system documentation is situated ....
....

Figure 6.2   Model certificate of testing of a fire alarm system

---

Certificate of testing of fire alarm system at:

Protected area ....

Name of occupier ....

Address ....

....

The system is operational and has been checked and tested in accordance with
BS 5839: Part 1: 1988

* clause 27	Extensions and alterations to an existing system
* subclause 29.2.6	Quarterly inspection and test
* subclause 29.2.7	Annual inspection and test
* subclause 29.2.8	Wiring check
* subclause 29.3.2	Servicing after a fire
* subclause 29.3.3	Servicing following a false alarm
* subclause 29.3.3	Servicing following excessive false alarms
* subclause 29.3.4	Servicing following a fault
* subclause 29.3.5	Servicing following a pre-alarm warning
* subclause 29.3.7	Other non-routine attention (specify)

....

....

* Delete if not applicable

Signed .... Status .... Date ....

Name (in block letters) ....

For (user or service organisation)..

....

Figure 6.3   Log book

Foreword

It is recommended that this log book is maintained by a responsible executive who should ensure that every entry is properly recorded. An 'event' should include fire alarms (whether real or false), faults, pre-alarm warnings, tests, temporary disconnections and the dates of installing or service engineer's visits with a brief note of work carried out and outstanding.

Reference data
Name and address ....

....

Responsible person ....Date ....

....Date......................

....Date ....

The system was installed by...

and is maintained under contract by ....until........................

Tel. no. .... who should be contacted if service is required.

Event data

Date	Time	Counter reading*	Event	Action required	Date completed	Initials

Expendable component replacement due (list): ....

....

....

*If an event counter is provided.
A program controlled system may need a column to record the readings of the "failure to correctly execute software" counter.

Chapter 7   TESTING

BS 7671 requires the following tests (including measurements where specified) to be carried out on fixed LV electrical installations:

> continuity
> insulation resistance
> polarity
> earth fault loop impedance
> earth electrode resistance
> operation of RCDs.

In this chapter these tests and measurements are discussed, plus insulation resistance tests on the HV windings of HV/LV transformers and on motors and generators.

(NOTE - INSTALLATIONS MUST BE ISOLATED BEFORE TESTING (except RCD and loop impedance))

For detailed guidance on testing electrical installations see IEE Guidance Note No. 3, Inspection and Testing.

7.1   Continuity

Amendment No 1 to BS 7671 introduced the requirement that continuity testing of conductors shall be carried out with an instrument having a no-load voltage of between 4 V and 24 V d.c. or a.c. and a short-circuit current of not less than 200 mA. This aligns the requirement of BS 7671 with the CENELEC Harmonisation Document (HD 384.6.61 S1).

It is to be noted that there are no particular merits in carrying out the measurement with an instrument that develops a minimum short-circuit current of 200 mA, as compared with one that develops less than 200 mA, currents of this order do not stress the conductor or connections.

There are two basic test methods:

1)   Applying a temporary shorting link at the fuseboard - see Figure 7.1a.

2)   Using a flying lead - see Figure 7.1b.

Where ferrous enclosures such as steel conduit or trunking are used as the protective conductor, particular care must be exercised when testing as a casual contact may give a satisfactory test result. It is recommended that the following procedure be followed:

1)   inspect the enclosure along its length, checking all the joints for electrical and mechanical soundness
2)   carry out a standard continuity (ohmmeter) test
3)   if the inspector has concerns regarding the soundness of the conductor a high-current test may be necessary; a.c. ohmmeters with 50 V output are available

that can be used to test at up to 1.5 times the design current up to a maximum of 25 A. Care needs to be exercised since sparking can occur at loose joints.

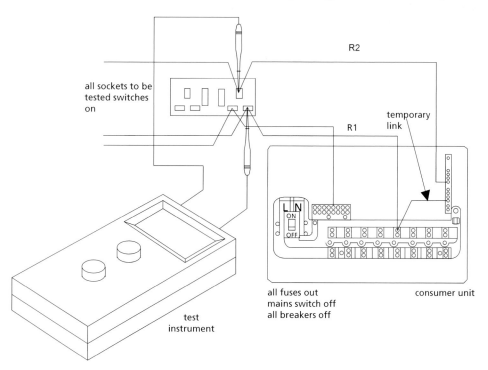

Figure 7.1a Testing continuity of protective conductors by shorting

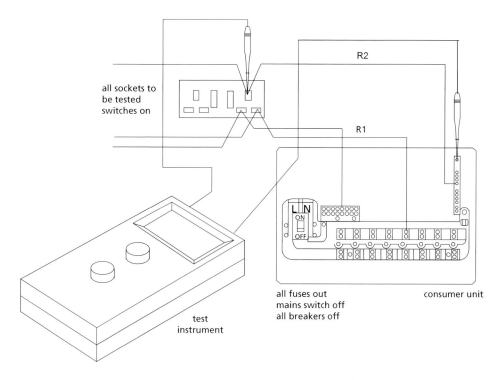

Figure 7.1b Testing continuity of protective conductors with a flying lead

7.2 Insulation Resistance

BS 7671

BS 7671 requires measurement for 230/400 V systems at 500 V d.c. and requires the instrument to be capable of supplying the test voltage when loaded with 1 mA. That is, the insulation tester must be able to maintain 500 V when loaded with the minimum acceptable insulation resistance of 0.5 MΩ. Instruments not able to meet this requirement generally can continue to be used except where insulation resistances are 0.5 MΩ or less.

Although an insulation resistance of not less than 0.5 MΩ complies with BS 7671, where a reading of less than 2 MΩ is recorded on a distribution board (with all circuits connected) the possibility of a defect exists. Then, each circuit should be individually tested and the measured insulation resistance of each should exceed 2 MΩ.

General

Electrical equipment cannot function properly if its electrical insulation is not in good condition. Generally, insulation deteriorates relatively slowly so that by regular monitoring action can be taken before the deterioration becomes critical.

The insulation resistance of materials is affected by many factors - moisture, contamination, oil, corrosive substances, vibration, heat and ageing. For insulators which can absorb water, the water content is probably the most important factor in determining insulation readings. Many of the effects on insulation are reversible so that insulation readings may depend upon the type of weather, whether the machine or equipment is at running temperature etc. The insulation of mineral insulated cooker elements will be relatively low if the elements have absorbed moisture. After, say, half an hour of use, the moisture will be driven off and the insulation reading will be considerably higher.

Flash Testing

The condition of the insulation of equipment is sometimes tested by an applied voltage test, often called a flash test or a proof test. Applied voltage tests differ significantly from insulation resistance measurement. An applied voltage test is performed only once on equipment after manufacture, or repair. The insulation either withstands the applied voltage test or it does not. If the insulation breaks down, damage is likely although many of the portable appliance test instruments on the market are so constructed that the current flowing on failure of the insulation is limited to prevent damage to the equipment. Repeated application of high voltage may damage insulation by creating voids or making existing voids larger.

Insulation resistance

Insulation resistance tests or measurements should be non-destructive. The value of the resistance is obtained by applying a voltage across the insulation and measuring the current flow. The current that passes through the insulation (or over it along a creepage path) is measured and, by Ohm's law, the resistance obtained.

Current and time

The measured resistance of insulation depends upon a number of factors, most particularly humidity and temperature. Other factors that particularly influence the reading are:

    1) the nature of the current - e.g. a.c. or d.c.
    2) duration of the test.

The total current that flows is often considered to be made up of three currents:

    1) capacitance charging current
    2) absorption current
    3) conduction or leakage current.

Figure 7.2   Time-current curves for d.c. insulation testing.

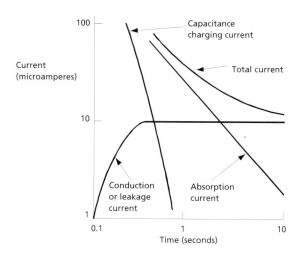

Capacitance charging current

For d.c. applied voltages, the capacitance charging current is initially high and drops as the insulation becomes charged up to the applied voltage. For a.c. applied voltages the capacitive current will be constant. As a general rule, a.c. tests will give lower insulation readings than d.c. tests.

Absorption current

The absorption current is initially high and reduces with time but at a much slower rate than the capacitance charging current.

Conduction current

The conduction or leakage current is generally small and steady.

The net effect of these currents when d.c. tests are applied, is for an initial low resistance reading building up to a steady reading after the order of 10 seconds.

Following an insulation test, energy is stored in the insulator. This is due to the capacitance charging current and the absorption current. To make the equipment safe after the test the stored energy must be discharged. Discharging of this energy will take longer than the initial charging, so sufficient time must be allowed for the discharge, at least 3 times as long as the time to charge.

Use of insulation testers

There are some very important do's and don'ts that apply to insulation measurements that are essential for safety:

1. equipment or circuits to be tested must be switched off and isolated and discharged
2. circuits or equipment must be discharged immediately after testing and before touching equipment or test leads.

On testers with voltage ranges, the discharge can be monitored and when a safe level is reached that is below 50 volts the equipment and test leads can be touched. Otherwise, a time period of four times that of the applied test should be allowed after carrying out measurements for dissipation of the stored energy.

Insulation tester with a guard terminal

Some insulation testers have a guard terminal. This terminal is used to obtain more accurate or more consistent readings of the insulation resistance.

When testing cables the guard terminal can negate the effect of surface leakage at cable ends. When testing equipment such as transformers, the guard can enable more specific measurements of the insulation resistance to be made. See Figures 7.4 and 7.5.

The current measured by the insulation resistance meter can be considered in two parts:

1) the current that flows through the body of the insulation
2) the current which tracks over the surface.

Whilst in terms of assessing the likely breakdown of a piece of equipment, it is important that both of the above currents are measured, there are circumstances where it is useful to eliminate the surface leakage. For example, if a cable is being tested prior to termination, it will be wished to eliminate the surface leakage at the cable ends from the measurement.

In Figure 7.3a the surface leakage is considered as two resistors.

Figure 7.3a Surface leakages considered as two resistors

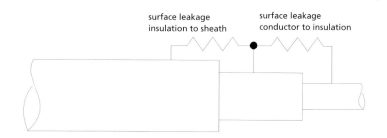

The guard is connected to, in effect, the centre of these two resistors with a view to eliminating as much as possible the surface leakage resistance from the meter reading.

The effective circuit is then as given in Figure 7.3b:

Figure 7.3b Resistances as seen by the insulation tester

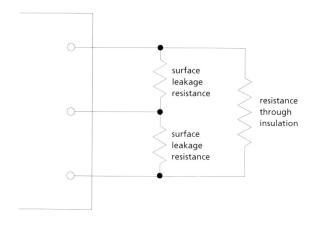

Some typical connection arrangements with the guard terminal are shown in Figure 7.4

Figure. 7.4a Guard connected to eliminate surface leakage over insulation at both ends of cable. Note:- one conductor is used to complete the guard circuit.

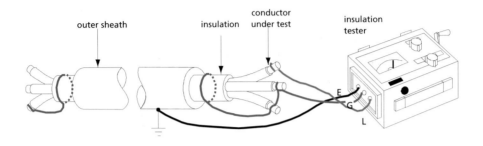

Figure 7.4b Guard connected to eliminate surface leakage over insulation and from adjacent conductors

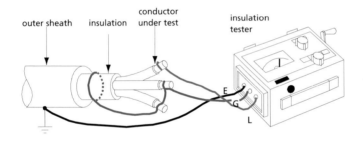

Figure 7.4c Guard used to eliminate leakage to other wires, when insulation resistance between one wire of a bunch and earth is being measured.

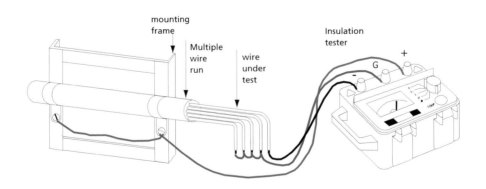

Equipment Measurements

When measuring the insulation resistance of equipment such as transformers, the use of the guard terminal can eliminate measurement of current through an unwanted element of the equipment.

Figures 7.5a and 7.5b show how measurements of insulation resistance of HV windings and bushings of transformers can be carried out, eliminating leakage across to the LV winding.

Figure 7.5a Connections for testing the insulation resistance of the HV winding and bushing of a transformer, eliminating leakage to the LV winding

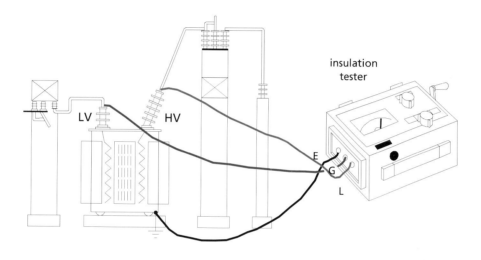

Figure 7.5b Connections for testing the insulation resistance between the windings of a transformer, eliminating leakage to earth

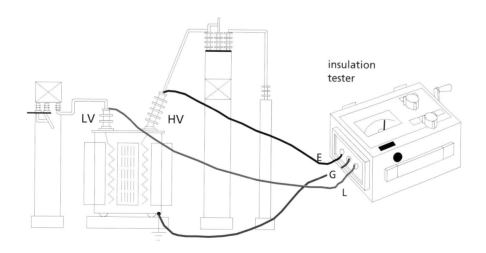

Short time or spot test

With this test method, the test is applied for a short but specified period of time, usually 60 seconds. The insulation resistance reading will increase with time as shown by Figure 7.6.

Fig. 7.6   Insulation resistance/time curve for short-time or spot-reading method

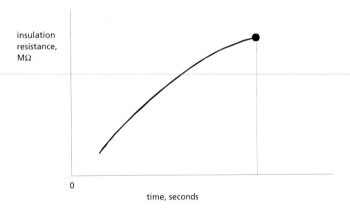

The usefulness of this test is that, if repeated at regular intervals, it will give an indication of the trend in the insulation condition. The trend is more marked with the short time test than it is with a long time test and so can give an earlier indication of deterioration.

Time-resistance or absorption tests

Insulation that is contaminated by dirt and moisture will not have as noticeable an absorption effect as good insulation. Readings are taken over a period of time, see Figure 7.7.

7.7   **Typical time-resistance curves**
curve a - shows absorption effect: insulation is satisfactory.
curve b - reveals contaminated insulation because resistance doesn't increase.

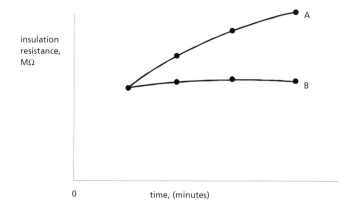

Curve A shows the typical results for good insulation and curve B shows contaminated insulation. The tests can often be used without reference to past results if the tester has experience of similar equipment. Because of the length of time that the test needs to be applied, it is not really practical with a hand-cranked generator.

Dielectric absorption ratio and polarisation index tests

The dielectric absorption and polarisation index test are in effect time-resistance tests, where the test instrument itself takes the measurements of time and current automatically and provides an output of the ratio of the readings at particular time intervals. The polarisation index readings are normally taken at 1 minute and 10 minutes. A comparison of the sort of readings that would be obtained is given in Table 7A below.

Table 7A Polarisation Index

Dielectric Absorption 60 sec/30 sec ratio	Polarisation Index 10 min/1 min ratio	Insulation Condition
---	<1	unsatisfactory
1,0 to 1,25	1,0 to 2,0	dubious
1,4 to 1,6	2,0 to 4,0	good
>1.6*	>4,0*	very good

*If values much higher than this are recorded, it may indicate dry brittle winding insulation which will break down under shock e.g. on start-up.

There are two recognised conditions that are not readily identified by PI tests. These are:

(i)   dry brittle insulation which gives a high PI reading

(ii)   multi-layer insulation where there is a failure of one
         of the layers.

In the case of (ii), such failed layered insulation may give a higher PI reading than layered insulation in good condition.

Dielectric discharge test

As mentioned above, the dielectric absorption ratio may not identify failure of insulation layers. The dielectric discharge test does assist in identifying such failure.

A layered insulation has an equivalent circuit to a number of high resistances shorted by capacitances, as shown by Figure 7.8.

Figure 7.8   Layered insulation

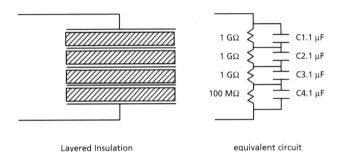

Layered Insulation   equivalent circuit

If one layer is damaged, it has less resistance than other layers but the same capacitance. A dielectric discharge test can identify such a failure. This test normally applies a steady d.c. voltage for a period such as 30 minutes. The entire insulator is then discharged quickly through a discharge resistor of approximately 0.5 MΩ and the discharge current is measured after a specific time such as 2 or 3 seconds. The dielectric discharge figure is given by:

DD = (current flowing after one minute) / (applied voltage x capacitance).

Typical DD test results are shown in Table 7B.

Table 7B   Dielectric discharge test results

DD Test Result	Insulation Condition
>5	Unsatisfactory
3 – 5	Dubious
1 – 3	Acceptable
<1	Satisfactory

DD test results can be taken over a wide range of current and capacitance conditions. A DD result will be produced if the capacitance is in the range 0.2 μF to 10 μF and reverse measured current does not exceed 10 μA.

Step voltage tests

Step voltage tests are carried out at two voltages, typically 500 V followed by tests at 2500 V, after allowing sufficient time for discharging of the test piece. If the insulation resistance reading obtained at the high voltage is less than at the lower voltage then a weakness is revealed which may require further investigation. The tests are basically a variation of the time-resistance method. After a short test time the readings at both voltages should be approximately the same; however, if long test times are used the lower test voltage will usually give a higher resistance readings. A difference in reading with the step voltage test is normally an indication of contamination of the test piece e.g. with moisture.

Motors and generators

Faulty rotating machines will normally have one of two causes:

      a)   an open circuit conductor
      b)   breakdown or partial breakdown of insulation.

Partial breakdown of insulation is quite common and results in such problems as armatures burning out, or arcing at brushes.

To avoid such breakdowns, which could be particularly costly in terms of lost production or disturbance to normal working, routine insulation resistance tests should be conducted and recorded in a log book to provide a historical record.

Note: Mechanical inspections are also necessary, checking for loose fixings, excessive play, or vibrations, irregular rotation, etc.

Test conditions

When testing machines it is important to record the test conditions.

The environment has a considerable effect on readings. If the weather is damp, considerably lower readings will be obtained for a machine than in dry conditions.

The effect of temperature on machines is generally different from that on such equipment as metal sheathed heating elements. The resistance of the insulation of motor and generator windings falls as temperature rises and therefore a machine which has been running hot is likely to give lower insulation readings than when cold. Weather conditions also affect insulation resistance, cold dry weather inclining towards high resistance readings. The resistance of a hot metal sheathed element will generally be higher when hot than when cold, as the mineral insulation will have dried out.

Typical readings for insulation tests on three power house generators performed at 500 V d.c. are given in Table 7C.

Table 7C   Typical Generator Insulation Resistances

Generator No. 1   40 kW.110 volts

Date	Insulation Resistance MΩ	Weather Conditions	Time of Test and Temperature of Machine
Feb.25	15	Fine and bright	Machine hot
27	10	-	Monday morning, (machine cold)
Mar.10	50	Cold and dry	Machine hot
12	30	Very cold and dry	Monday morning, (machine cold)

Generator No. 2   50 kW.110 volts

Feb.25	25	Fine and bright	Machine hot
27	15	Fine and bright-	Monday morning, (machine cold)
29	50	Dull and dry	Hot after 12 hours at 28 kW
Mar.1	12	Rainy	After standing 12 hours
10	100	Cold and dry	Machine hot
12	50	Very cold and dry	Monday morning, machine cold

Generator No. 3   135 kW.460 volts
New machine being erected.

Feb.24	20	-	Armature only
24	25	-	Field only
Mar. 5	4	-	Complete machine on completion of erection
8	2,5	Fine	Machine hot
9	30	Fine	Machine cold after standing all day
12	75	Very cold and dry	Machine cold
23	20	Fine	Machine hot Test after running all day

Practical tests of d.c. machines

The following procedure may be followed when testing machines:

1)   Open the main switch, remove fuses and isolate the entire motor or generator from the supply;

2)   Connect together both terminals on the machine side of the isolator, connect these to the L or negative terminal of the insulation tester and connect the E or positive terminal to earth via the motor frame earth terminal or switch case;

3)   Operate the insulation tester and take a resistance reading.

     If the testing at the main switch provides an unsatisfactory reading, the next step should be :-

4)   Disconnect all cables at the machine and re-test - if the reading remains unsatisfactory the fault is on the machine. If satisfactory, the fault is on the controlgear or the cabling.

The following tests are required to locate a machine fault:

First, lift all brushes to separate the armature from the main field winding (check also interpoles and compensating field windings).

1)   Motor frame to each winding and brush gear;

2)   Motor frame to separated field windings;

3)   Motor frame to armature winding.

7.3   Polarity

Polarity checking is carried out to confirm that all equipment is connected as necessary to the correct pole of the supply; for example, that switches and fuses will interrupt the phase and not the neutral of the supply. There is a requirement in BS 7671 to carry out continuity testing of an installation before the supply is connected and this can be used to confirm that the polarity is correct and that all switches and fuses are in the phase conductor(s).

There are instruments that give an indication of polarity when the supply is live. The most common are test lamps and neon testers. It is to be noted that test lamps and neon indicators simply indicate that there is a potential difference. If there is concern that the polarity of a PME (TN-C) supply coming into the premises is incorrect, the test of polarity will need to be made with a neon indicator as per Figure 7.9 with the main fuse removed and the neutral/earthing conductor disconnected from the main bonding. Tests between phase (L), neutral (N) and earth (E) terminals will not identify the cross. The test must be to true earth, not the earth terminal, and is most easily carried out with a neon indicator with main bonding disconnected.

Figure 7.9   Testing polarity of TN-C supplies

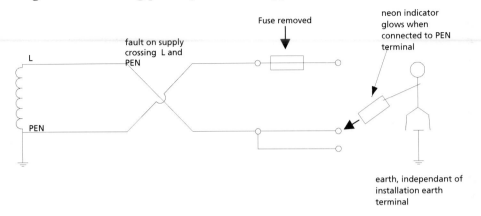

Many phase-earth loop testers and RCCB testers are fitted with neon lights to check for correct wiring in an installation prior to measurement of loop impedance and RCD tests.

The indication is normally by neon again initiated by a sufficient voltage between the phase and neutral/earth of the installation. Again, these checks indicate a potential difference and they will not necessarily indicate that there is a neutral earth connection reversed or indicate incorrect polarity of the incoming supply.

7.4   Earth fault loop impedance

Earth fault loop impedance instruments should only be used for the specific test application for which they were designed. Typically, an earth fault loop impedance measuring instrument determines impedance by charging a capacitor on a first half cycle and discharging that capacitance on the second half of the cycle with a resistance in series. The charge remaining is used to calculate the loop impedance.

Some of these instruments may be also used to measure fault level. This is the single-phase to earth fault level. For three-phase installations an approximation of the fault level can be obtained by doubling the single-phase to earth reading.

Because of the nature of the measurement, loop impedance testers may give misleading readings if used for any tests other than a standard loop test, or earth fault current measurement.

7.5   Earth electrode test

For an absolute measurement of the resistance to earth of an earth electrode, the measurement preferably should be carried out using a specialist earth electrode resistance tester, which may be a four-terminal tester as illustrated in Figure 7.10, or a three-terminal tester, in which the functions of the C1 and P1 terminals are combined. Such testers must be used when measuring earth electrodes for generators and distribution system earthing, such as at 11 000/415V transformers.

Figure 7.10 Earth electrode resistance measurement

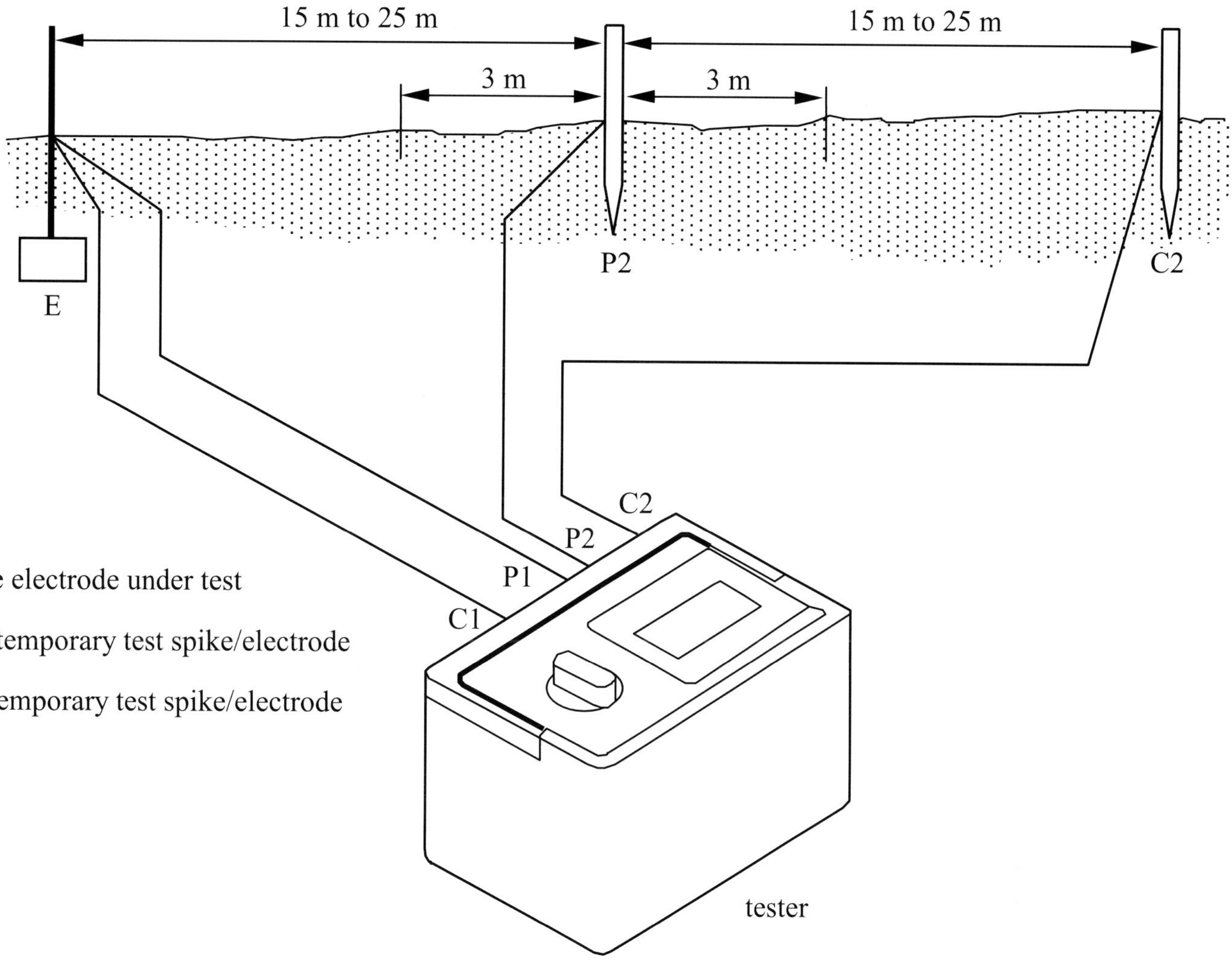

where:

E   is the electrode under test

C2   is a temporary test spike/electrode

P2   is a temporary test spike/electrode

AFTER COMPLETION OF THE TESTING ENSURE
THAT THE EARTHING CONDUCTOR IS RECONNECTED

For measuring the resistance to earth of earth electrodes installed in domestic installations for use with an RCD, an earth fault loop impedance tester may be used. Before the test is undertaken all equipotential bonding should be disconnected from the earth electrode under test, to ensure that the test current passes through the earth electrode alone. The bonding must be reconnected after the test.

The loop tester is connected between the phase conductor at the origin of the installation and the earth electrode with the test link open. The impedance reading obtained is that of the complete phase earth loop including the resistance of the earth electrode. As the earth electrode resistance will be the major element of the resistance of this loop, a reading of sufficient accuracy for these circumstances is given, which errs on the high side.

7.6 Infra-red cameras

Infra-red cameras give a measure of the temperature of electrical components and can identify hot spots in equipment and installations. This gives an indication of the condition of equipment, and in particular of terminations that might be loose or about to fail.

The particular advantage of such cameras is that equipment can be quite safely scanned with the camera without disconnecting the supply.

7.7 Instrument standards

The safety-in-use standard is BS EN 61010 : Safety requirements for electrical equipment for measurement, control and laboratory use. The performance standard, which also requires compliance with BS EN 61010, is BS EN 61557 : Electrical safety in low voltage distribution systems up to 1000 V a.c. and 15 00 V d.c. Equipment for testing, measuring or monitoring of protective measures. The parts published include :

BS EN 61557-1: 1997 - General requirements
BS EN 61557-2: 1997 - Insulation resistance
BS EN 61557-3: 1997 - Loop impedance
BS EN 61557-4: 1997 - Resistance of earth connection and equipotential bonding
BS EN 61557-5: 1997 - Resistance to earth
BS EN 61557-7: 1997 - Phase sequence
BS EN 61557-8: 1997 - Insulation monitoring devices for IT systems.

Chapter 8   INDUSTRIAL SWITCHGEAR MAINTENANCE

8.1   Switchgear register

Before carrying out any work on industrial switchgear, including operation or routine maintenance, a register should be prepared to identify manufacturer, model, type, maintenance and modification history. Then, with knowledge of the maximum prospective fault current at each location and the load at each location, it may be determined if the switchgear is potentially fit for the purpose and if, after maintenance and if necessary repair, it may be safely operated. A typical register is shown in Figure 8.1.

At each location the maximum prospective fault current must be determined, and as advised in BS 7671 this can be measured, calculated or ascertained. Where fault levels are measured using a single-phase device, it must be remembered that the maximum prospective three-phase fault level is likely to be at least twice the single-phase reading. This is because, for a three-phase to earth fault, the neutral or earth return path will have no influence. The impedance of the earth return needs to be removed from the calculations of fault current. For fault levels above 20 kA, specialised equipment must be used or the fault level determined by calculation, or by enquiry of the supply company.

It is preferable to request the fault level at the origin of the installation from the electricity supply company as any instantaneous measurement may not represent the maximum prospective fault level. The maximum prospective fault level will depend upon the configuration of the supply company's network. If fault levels are calculated the network configuration giving the highest fault level must be presumed.

Fault ratings may be obtained from the equipment rating plate, or by enquiry of the manufacturer (the serial number will need to be quoted). If the fault rating of the switchgear does not exceed that of the fault level, the switchgear must be replaced. If the switch is oil-filled particular care must be taken, as described in the next section.

8.2   Oil-filled switchgear (HV and LV)

If oil-filled switchgear (both LV and HV) is installed a copy of the Health and Safety Executive information document 483/27 should be obtained. This document provides information for managers and technical staff, concerning the risks that can arise from the use of high voltage and low voltage oil-filled electrical distribution switchgear manufactured prior to 1970. Advice is given on precautions which should be taken to eliminate or control these risks.

Oil-filled switchgear has been made for sixty years and if manufactured to current standards, and when maintained and operated correctly, represents no risk. However, should there be a fault on oil-filled switchgear there could be an explosion of burning oil, which not only provides a risk of fire to the premises but also a very real risk of death or serious injury to anyone who might be operating or close to the switchgear. For these reasons it is particularly important to carry out thorough investigations into the maintenance and modification history of any oil-filled switchgear installed. Responsible persons have obligations set out in the

Figure 8.1 Switchgear and transformer register

Switchgear record

Substation name / no ....

Location ....

Switch no	
Make	
Year of manufacture	
Year of commissioning	
Year of modification	
Gear type	
Serial number	
Duty	
Type	
Fault Rating MVA	
Load Rating A	
Protection C ratio	
Protection V ratio	
Operating mechanism	

Transformer Record

Substation name / no ....

Location ....

Transformer no.	
Make	
Year of manufacture	
Year of commissioning	
Rating	
Type	
Serial no.	
Vector Group	
Primary kV / Secondary V	
Tap range	
Tap steps	
Percentage impedance	
Oil capacity (litres)	
Breather	
Conservator	
Weight	

Health and Safety at Work etc. Act, the Management of Health and Safety at Work Regulations and the Electricity at Work Regulations.

Staff carrying out maintenance work on an installation including oil-filled switchgear, either low voltage or high voltage, should be aware of particular aspects concerning its operation:

1. Knowledge is required of the particular switchgear installed and experience of its operation.
2. The fault rating of some switchgear is low and may not be suitable for the fault level of the installation.
3. Such switchgear may have required, on the advice of a manufacturer, modifications and the maintainer will need to know if such modifications were necessary and if they were carried out.
4. Older switchgear may be manually operated, the movement of the switchgear contacts being directly dependent upon the movement of the handle by the operator. Such switchgear will need replacement or modification.
5. Oil-filled switchgear needs routine maintenance, including checking of the oil condition and checks for oil leaks, condition of switchgear contacts etc.
6. HV oil switches or isolators should be fitted with anti-reflex operating handles.

It is not appropriate to go into detail on these matters here as maintenance staff could be misled into thinking by reading this guidance that they will gain sufficient knowledge and experience to carry out the work. The operation of inadequate or defective switchgear is very dangerous and has resulted in serious burns and fatalities.

If such switchgear is found it must be fully identified, including noting the following:

(1) location;

(2) manufacturer and type reference for each item of equipment and type of equipment;

(3) serial number and year of manufacture;

(4) date of installation;

(5) voltage rating;

(6) current rating;

(7) fault rating and whether it is certified or assessed rating;

(8) type of operating mechanism (dependent manual, independent manual, dependent power, independent power or stored energy);

(9)   details of any modifications or repairs, e.g. fitted anti-reflex handles;

(10)   date equipment last maintained/serviced;

(11)   if the equipment is an oil circuit-breaker whether it is plain break equipment (i.e. equipment without arc control devices) or not; and

(12)   type of electrical protection fitted and details of the settings, e.g. 300/5 A c.c.s., time-lag fuses, direct acting trips.

It is then necessary to contact competent persons for a full investigation to be carried out. It may well be that the switchgear is in suitable order and after routine maintenance can be used. However, this may not be. It is then important to take precautions to reduce the risk, pending modification or replacement. If the switchgear does not have sufficient fault rating, all live operation and automatic tripping of the switchgear should be prohibited and prevented. All access to the switchgear when live should be prevented.

It is stressed here that failure of such switchgear can result in fatalities.

8.3   Routine maintenance of high voltage equipment

High voltage switchgear and equipment such as transformers requires regular routine maintenance. This work will need to be entrusted to a competent organisation. This guidance cannot provide all the information necessary with respect to HV maintenance, but only an outline of the work that is likely to be required is given to assist in placing orders for the work.

Generally maintenance is considered as three activities: inspection, examination and overhaul. The recommended maximum periods between these maintenance activities are given in Table 8A.

Table 8A   Frequency of maintenance for 11000 V switchgear and 11000/415 V transformers

Maintenance Activity	Maximum period between maintenance activities
Inspection	1 year
Examination	5 years
Overhaul	15 years

It is important to note that there may be circumstances where the frequency of all the maintenance activities should be reduced due to unfavourable conditions such as frequent operation of the switchgear or an onerous environment.

It is important to note that all oil circuit-breakers must be overhauled as soon as possible after fault operation and SF$_6$ circuit-breakers will need to be examined as soon as possible after fault operation.

Typical maintenance schedules are shown here. A person placing an order for such maintenance should expect all these activities to be covered by inspection, examination and overhaul documentation.

Oil Circuit-breaker Maintenance Schedule

Maintenance Operations	Inspection	Examination	Overhaul
General inspection	✓	✓	✓
Inspect time limit fuses	✓	✓	✓
Check settings of protective relays	✓	✓	✓
Cleaning		✓	✓
Trip and closing mechanism		✓	✓
Contacts (main and arcing)		✓	✓
Trunk or truck winding mechanism		✓	✓
Shutters	✓	✓	✓
Interlocks	✓	✓	✓
Bushings		✓	✓
Auxiliary contacts		✓	✓
Secondary wiring and fuses		✓	✓
Insulation test		✓	✓
Earth connections		✓	✓
Heaters		✓	✓
Insulating oil		✓	✓
Main and arcing contacts		✓	✓
Arc control devices		✓	✓
Switchgear spouts			✓
Venting and gas seals			✓
Tank and tank linings			✓
Operational check	✓	✓	✓
Voltage and current transformers			✓

Oil Switch Maintenance Schedule

Maintenance Operations	Inspection	Examination	Overhaul
General inspection	✓	✓	✓
Cleaning*		✓	✓
Withdrawal locking mechanism*	✓	✓	✓
Safety shutters*	✓	✓	✓
Insulators/bushings*			✓
Insulating oil			✓
Isolating contacts*			✓
Operating mechanism			✓
Main and arcing contacts			✓
Arc control devices			✓
High voltage fuse connections			✓
Switchgear spouts			✓
Tank and tank linings			✓
Interlocks	✓		✓
Auxiliary contacts			✓
Earth connections			✓
Weather shields			✓
Insulation test			✓
Operational check*		✓	✓

* Withdrawable types only

Maintenance Schedule for SF6 Switchgear

Maintenance Operations	Inspection	Examination	Overhaul
General inspection	✓	✓	✓
Cleaning		✓	✓
Leak tests		✓	✓
Lubrication		✓	✓
Operational check	✓	✓	✓

Oil-filled Transformer Maintenance Schedule

Maintenance Operations	Inspection	Examination	Overhaul
General inspection of exterior	✓	✓	✓
Oil level	✓	✓	✓
Breather	✓	✓	✓
Sampling and testing of suspect oil only	✓		
Sampling and testing of oil as fixed routines		✓	✓
Clean exterior			✓
Conservator, clean and inspect			✓
Check external connections and conductors			✓
Main tank			✓
Visual inspection of transformer core			✓
Check LV and HV connections			✓

Chapter 9   ELECTRONIC DATA PROCESSING INSTALLATIONS

9.1   Introduction

The British Standard Code of Practice for fire protection for electronic data processing installations is BS 6266: 1992. This Publication provides guidance on the selection of fire detection, fire alarm and fire extinguishing equipment. The recommendations with respect to electrical equipment are somewhat limited and are summarised here.

9.2   Electrical installation

Cables

The general requirement is that the electrical installation should comply with BS 7671 and that all cables should meet the minimum performance requirement of BS 4066 "Tests on Electric Cables under Fire Conditions", Part I : method of test on single vertical insulated wire or cable. Standard PVC and thermosetting cables in common use listed in BS 7671 meet these requirements.

It is recommended that, in ceiling and floor voids, power cables other than steel-wire armoured or mineral insulated cables should be installed in metal conduit or metal trunking. Data cables should also be contained within metal conduit or trunking or clipped to a metal tray.

There is a further general recommendation that all cables should be specified as having low emission of smoke and corrosive gases when affected by fire i.e. meeting the requirements of BS 7622 : Measurements of smoke density of electric cables burning under defined conditions and BS 6425 : Gases evolved during combustion of electric cables.

To assist in the selection of cables with these properties, Table 7.3 in IEE Guidance Note 4 Protection against Fire is reproduced here as Table 9A.

Table 9A   Fire-related properties of cable standards

Cable Standard		Fire related properties	
BS 6207	Mineral insulated cables with a rated voltage not exceeding 750 V	BS 4066 Pt 1 & Pt 2 Cat A	Tests on electric cables under fire conditions
		BS 7622 (for cables with zero-halogen coverings)	Measurement of smoke density of electric cables burning under defined conditions
		BS 6425 (for cables with zero-halogen coverings)	Gases evolved during combustion of electric cables
		BS 6387 Cat C, W & Z	Performance requirement for cables required to maintain integrity under fire conditions
BS 6724	Armoured cables for electricity supply having thermosetting insulation with low emission of smoke and corrosive gases when affected by fire	BS 4066 Pt 1 & Pt 3 Cat C	Tests on electric cables under fire conditions
		BS 7622 (Currently BS 6724 Appendix F)	Measurement of smoke density of electric cables burning under defined conditions
		BS 6425	Gases evolved during combustion of electric cables
BS 7211	Thermosetting insulated cables (non-armoured) for electric power and lighting with low emission of smoke and corrosive gases when affected by fire	BS 4066 Pt 1 or Pt 2 as specified & Pt 3 Cat C (two to five-core circular 450/750 V)	Tests on electric cables under fire conditions
		BS 7622	Measurement of smoke density of electric cables burning under defined conditions
		BS 6425	Gases evolved during combustion of electric cables
BS 7629	Thermosetting insulated cables with limited circuit integrity when affected by fire	BS 4066 Pt 1	Tests on electric cables under fire conditions
		BS 7622	Measurement of smoke density of electric cables burning under defined conditions
		BS 6387 Cat B, W & X	Performance requirement for cables required to maintain integrity under fire conditions
		BS 6425	Gases evolved during combustion of electric cables
BS 7846	600/1000 V armoured fire resistant electric cables having low emissions of smoke and corrosive gases when affected by fire	BS 4066, Pt 1 & Pt 3 Cat C	Tests on electric cables under fire conditions
		BS 7622 (Currently BS 7846 Appendix F)	Measurement of smoke density of electric cables burning under defined conditions
		BS 6425	Gases evolved during combustion of electric cables
		BS 6387 Cat C	Performance requirement for cables required to maintain integrity under fire conditions

There are also cable requirements specified in a number of system standards associated with electrical installations and these are reproduced in Table 9B (is Table 7.4 in IEE Guidance Note 4).

Table 9B   Cable specifications of system standards

Standard		Example Requirement
BS 5266 Part 1	Emergency Lighting Code of practice for the emergency lighting of premises other than cinemas and certain other specified premises used for entertainment	In general, the following cables to be used for connecting luminaires to the power supply, 8.2.2(a) those with inherently high resistance to attack by fire: (a) micc cable to BS 6207 (b) cables to BS 6387 category B e.g. BS 7629 and BS 7846 Cables as follows require additional protection 8.2.2(b): (c) pvc cables to BS 6004 in steel conduit (d) pvc cables to BS 6004 in rigid pvc conduit, classification 405/100,000 or 425/100,000 to BS 6099 Section 2.2 (e) pvc or XLPE insulated and sheathed steel-wire armoured cables to BS 6346 or BS 5467.
BS 5839 Part 1	Fire detection and alarm systems for buildings Code of practice for system design installation and servicing	Cables required to operate during exposure to fire (17.4.2) (a) micc cable to BS 6207 (b) cables to BS 6387 category AWX or SWX, e.g. BS 7629 and BS 7846. Other cables may be used provided they are protected against direct exposure to fire (see 17.4.2).
BS 6266	Fire protection for electronic data processing installations	Cables to comply with BS 4066 Part 1, e.g. all pvc cables. Where appropriate power cables should comply with BS 6724 (armoured thermosetting low smoke and gas) 4.2.1.5. In critical installations power cables to meet category AWZ of BS 6387 (note to 4.2.3.5) e.g. micc to BS 6207 and cables to BS 7629.

Emergency switches

It is recommended that emergency switches with manual, not automatic, resetting should be installed near to the electronic data processing equipment control console or at suitably located positions. Suitably labelled emergency switches should be placed near exit doors or at other suitable points to cut off power supplies to:

a)   air conditioning
b)   main power excepting lighting and smoke extraction.

A main isolator controlling all power supplies to the electronic data processing area except lighting and smoke extraction and ventilation is recommended to be provided at or near to the main entrance to the EDP area and marked:

FIRE EMERGENCY SWITCH

It is recommended that the number of junction boxes in underfloor areas should be kept to the minimum; they should be metal completely enclosed and readily accessible. It is also recommended that within floor voids no electrical equipment other than cables, junction boxes and smoke detectors should be installed.

The power cables should be routed around the perimeter of the data processing room to reduce interference and it is recommended that they should have no joints.

Chapter 10   RISK ASSESSMENT, MANUAL HANDLING, DISPLAY SCREEN REGULATIONS

10.1   Risk assessment

The Management of Health and Safety at Work Regulations impose duties with respect to the assessment of risks upon every employer and self-employed person. The particular regulations are as follows (with underlining added):

Regulation 3(1)
(1)   Every employer shall make a suitable and sufficient assessment of -
(a)   the risks to the health and safety of his employees to which they are exposed whilst they are at work; and
(b)   the risks to the health and safety of persons not in his employment arising out of or in connection with the conduct by him of his undertaking, for the purposes of identifying the measures he needs to take to comply with the requirements and prohibitions imposed upon him by or under the relevant statutory provisions.

Regulation 4(1)
Every employer shall make and give effect to such arrangements as are appropriate, having regard to the nature of his activities and the size of his undertaking, for the effective planning, organisation, control, monitoring and review of the preventive and protective measures.

Regulation 5
Every employer shall ensure that his employees are provided with such health surveillance as is appropriate having regard to the risks to their health and safety which are identified by the assessment.

Regulation 6(1)
Every employer shall, subject to paragraphs (6) and (7), appoint one or more competent persons to assist him in undertaking the measures he needs to take to comply with the requirements and prohibitions imposed upon him by or under the relevant statutory provisions.

Regulation 7(1)
Every employer shall establish and where necessary give effect to appropriate procedures to be followed in the event of serious and imminent danger to persons at work in his undertaking.

Regulation 5 introduces the concept of health surveillance as is appropriate, having regard to the risks identified by an assessment. Regulation 6 requires that the employer shall appoint one or more competent persons to assist in his understanding of the measures he needs to take. The first step any such competent person has to perform is to carry out an assessment of the risks. This applies to all work activities including maintenance.

Regulation 4 - Health and safety arrangements

Arrangements, perhaps procedures, need to be implemented for the planning, organisation, control, monitoring and review of preventive and protective measures. This applies to all work activities including maintenance. The regulation lists the sequence of events.

> Planning
> Organisation
> Control
> Monitoring and review

Regulation 5 - Health surveillance
Where appropriate, employers are required to implement health surveillance procedures where there are risks to their employees. These procedures may be necessary where there are risks of exposure to chemicals, radiation, extreme temperatures, infections, etc. In effect, in any situations where the human body's normal defence mechanisms may not be appropriate.

Regulation 6 – Competent persons
This requires the appointment of one or more competent persons to assist in complying with the regulations. This may mean that in particular instances specialists will have to be specifically employed to advise on the risks and the action to be taken.

Regulation 7 - Procedures
Where there are particular risks presenting serious and imminent danger such as fire, explosion, radiation etc. procedures must be established. Examples might include :

> Fire
> Leakage of hazardous substances
> Bomb alarms
> Accidents.

Risk assessment

It is important that the persons carrying out risk assessment are competent to do the work. Individuals who consider themselves not competent to carry out the risk assessments entrusted to them must advise their employer.

There can be considered six steps to risk assessment:

i)   identification of the hazard
ii)   identification of who might be harmed
iii)   evaluation of existing precautions
iv)   recording of findings - the hazard, who might be harmed, adequacy of existing precautions and recommendations for reduction of risks
v)   implementation of recommendations
vi)   review of assessment.

A pro-forma for such an assessment is shown in Figure 10.1.

Figure 10.1

RISK ASSESSMENT	Location	Assessment by
		Date of assessment
		Review recommended

THE HAZARD (List the hazards) Note 1	THOSE AT RISK (List the persons in groups at risk) Note 2	EVALUATION OF CONTROLS (List existing controls/procedures) Note 3	FINDINGS (List risks not adequately controlled) Note 4	Implementation of Findings — Record action taken	
				Date	Action

Notes to risk the assessment form Figure 10.1

1. Examples of common hazards include slipping/tripping, fire, chemicals, moving parts (e.g. of machines), work height, ejection of material, (steam, mouldings etc.) pressurised systems (e.g. boilers), vehicles (cars, lorries, fork-lift trucks), from wiring, dusts (including explosive dusts), fumes, gases, manual handling (lifting), noise, lighting levels, temperatures (hot or cold).

2. Consider not only those undertaking the work, but others who might be affected by the work activity.

3. Have adequate precautions already been taken? Consider information (signs, instructions), training, system procedures, barriers, guards, clothing (protective). Consider any legal requirements, industry and British or European Standards or codes of practice. Has the risk been reduced as far as is practicable? When risks are adequately controlled, the precautions taken need to be identified in terms of company procedures etc.

4. Summarise the actions recommended for risks not adequately controlled. This may need supplementing by a detailed description. Risk reduction must be prioritised on the basis of the hazard and the number of persons affected.

    Actions to be taken might include:

    remove the risk
    reduce the risk
    prevent access
    provide protection (e.g. clothing, glasses)
    provide monitoring.

10.2 Manual handling

The Manual Handling Operations Regulations 1992 came into force on 1 January 1993. They were made under the Health and Safety at Work etc. Act 1974 and implemented European Directive 90/269/EEC. The Health and Safety Executive publish guidance on these regulations called Manual Handling Reference L23.

Regulation 4(1) is reproduced below.

Each employer shall:

> a)   so far as is reasonably practicable, avoid the need for employees to undertake any manual handling operations at work which involve the risk of their being injured;

b)   where it is not reasonably practicable to avoid the need for employees to undertake any manual handling operations at work which involve the risk of their being injured.

    i) make a suitable and sufficient assessment of all such manual handling operations to be undertaken by employees having regard to the factors as specified in column 1 of Schedule 1 of these Regulations and considering the questions which are specified in the corresponding entry in column 2 of that Schedule;

    ii) take appropriate steps to reduce to the reasonable minimum the risk of injury to those employees arising out of their undertaking any such manual handling operations.

The factors to which an employer must have regard and the questions which he must consider in making an assessment are reproduced in Figure 10.2, reproduced from Schedule 1 of the Manual Handling Operations Regulations.

Figure 10.2 Schedule 1 of the Manual Handling Operations Regulations

Factors to which the employer must have regard and questions he must consider when making an assessment of manual handling operations

Schedule	Regulation 4(1)(b)(i)	
	Column 1 *Factors*	Column 2 *Questions*

	1 The tasks	Do they involve:
		- holding or manipulating loads at distance from trunk?
		- unsatisfactory bodily movement or posture, especially:
		  - twisting the trunk?
		  - stooping?
		  - reaching upwards?
		- excessive movement of loads, especially:
		  - excessive lifting or lowering distances?
		  - excessive carrying distances?
		- excessive pushing or pulling of loads?
		- risk of sudden movement of loads?
		- frequent or prolonged physical effort?
		- insufficient rest or recovery periods?
		- a rate of work imposed by a process?

	2 The loads	Are they: - heavy? - bulky or unwieldy? - difficult to grasp? - unstable, or with contents likely to shift? - sharp, hot or otherwise potentially damaging?
	3 The working environment	Are there: - space constraints preventing good posture? - uneven, slippery or unstable floors? - variations in level of floors or work surfaces? - extremes of temperature or humidity? - conditions causing ventilation problems or gusts of wind? - poor lighting conditions?

	4 Individually capability	Does the job:

4 Individually capability — Does the job:

- require unusual strength, height, etc.?
- create a hazard to those who might reasonably be considered to be pregnant or to have a health problem?
- require special information or training for its safe performance?

5 Other factors — Is movement or posture hindered by personal protective equipment or by clothing?

10.3 Display screen regulations

The Health and Safety (Display Screen Equipment) Regulations require a risk assessment to be carried out by the employer at each visual display unit work station used in the purpose of the business or provided by him and used by operators. Guidance on the Regulations is provided by the Health and Safety Executive in its publication L26 Display Screen Equipment at Work. It is not appropriate here to look into these regulations in any detail as any person responsible for assessment will need to obtain a copy of HSE document L26.

Those responsible for assessments must be familiar with the requirements of the regulations and be able to:

1. assess the risks associated with the work station;
2. make a clear record of the assessment and communicate the findings to those who need to take action;
3. assess any action necessary;
4. recognise their own limitations and, as necessary, call upon further expertise.

It is necessary to review the assessments from time to time, particularly if there has been:

a major change in work station furniture
a major change to the hardware
a major change to the software
changes in task requirements.

It will also be necessary to review an assessment if the work station is removed or there are significant changes in the environment such as in the level of lighting, either natural or artificial.

Chapter 11 SAFETY SIGNS AND SIGNALS

11.1 Introduction

The Health and Safety (Safety Signs and Signals) Regulations 1996 implemented European Council Directive 92/58/EEC setting minimum requirements for the provision of safety signs at work. Following the risk assessment carried out as required by the Management of Health and Safety at Work Regulations (discussed in Chapter 10), all necessary safety signs and signalling procedures must comply with the Health and Safety (Safety Signs and Signals) Regulations. The Health and Safety Executive publish guidance on these regulations "Safety Signs and Signals - Publication L64". The emergency escape and first aid signs are shown in Chapter 5 of this publication, as are the fire fighting signs.

11.2 Sign colours

The regulations require specific colours for particular signs. Table 11A is reproduced from Schedule 1 of the regulations:

Table 11A   Colours of safety signs

Colour	*Meaning or purpose*	*Instructions and information*
Red	Prohibition sign	Dangerous behaviour
	Danger alarm	Stop, shutdown, emergency cut-out devices Evacuate
	Fire fighting equipment	Identification and location
Yellow or Amber	Warning sign	Be careful, take precautions Examine
Blue	Mandatory sign	Specific behaviour or action Wear personal protective equipment
Green	Emergency escape, first aid sign	Doors, exits, routes, equipment, facilities
	No danger	Return to normal

These colour codes are shown pictorially for prohibition, warning, mandatory and emergency escape signs in Figure in 11.1.

Guidance on the format for a wider range of signs is given in HSE document L64.

11.3   Signs

Signs on containers and pipes

The regulations have minimum requirements with respect to signs on containers and pipes. The labels on such containers or pipes should be in accordance with Directive 67/584/EEC and 88/379/EEC or as per the warning signs in the general descriptions. Such warning signs can be supplemented by additional information. When the containers are being transported they need to be supplemented by suitable signs. All signs must be mounted on the visible side in non-pliable self-adhesive or painted form. Labels used on pipes must be positioned visibly in the vicinity of the most dangerous points such as valves and joints and at reasonable intervals. Areas, rooms or enclosures used for the storage of dangerous substances or preparations must be marked and labelled.

Fire fighting equipment

Signs are specified for fire fighting equipment and these have been reproduced in Chapter 5.

Hand signals

The Regulations recognise and have requirements for acoustic signals, verbal communications and hand signals. There are clearly considerable advantages in all persons in all industries adopting the same general signs and signals.

Exit signs

Old-style text only signs (to BS 2560) are required to be replaced or supplemented by new signs in accordance with the regulations. Signs to earlier BS 5499 are considered to meet the requirements of the regulations. New signs should comply with the regulations format and the current BS 5499.

Because the Directive format is substantially different to the old-style EXIT signs, BS 5499 recommends a conversion for existing signs retaining the word exit and recommends that the new sign is for a period supplemented by arrow and exit signs, see Figure 11.1.

Figure 11.1 Safety signs

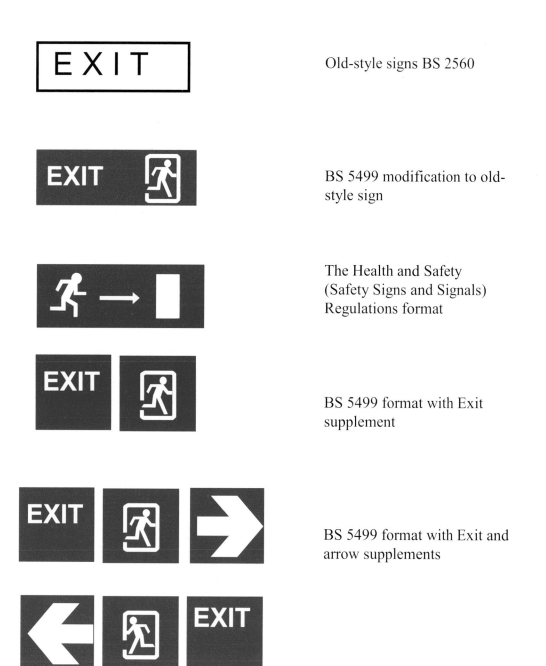

Old-style signs BS 2560

BS 5499 modification to old-style sign

The Health and Safety (Safety Signs and Signals) Regulations format

BS 5499 format with Exit supplement

BS 5499 format with Exit and arrow supplements

(i)   *Prohibition sign*
         Red/White/Black - a sign
         prohibiting behaviour likely
         to increase or cause danger
         (e.g. no smoking);

(ii)   *Warning sign*
         Yellow/Black - a sign giving
         warning of a hazard or
         danger (e.g. danger:
         electricity);

(iii)   *Mandatory sign*
         Blue/white - a sign
         prescribing specific
         behaviour (e.g. eye
         protection must be worn);

Fire fighting signs (Red and White)

Intrinsic features:

(a)   rectangular or square shape

(b)   white pictogram on a red background (the red part to take up at least 50% of the area of the sign).

Supplementary Green/White : 'This way' signs for fire fighting equipment

11.4   Identification and notices

The general requirements of BS 7671 for identification and notices are important to those responsible for the operation and maintenance of an electrical installation. Except where there is no possibility of confusion, a label or other means of identification is required to be provided to indicate the purpose of each item of switchgear and controlgear. This requirement is reinforced by the Electricity at Work Regulations 1989, and is essential for safe operation. Regulation 514-01-02 of BS 7671 goes further, and states that, as far as is reasonably practicable, wiring shall be arranged or marked so that it can be identified for inspection, testing, repair or alteration of the installation. This means that, for example, cable runs from switchboards should be grouped together, with cables and conductors running neatly and logically on cable trays etc.

Conduit and other pipes

Where conduit is to be distinguished from pipes or other services, orange is the colour to be used in accordance with code of practice BS 1710. The colour identifiers for other services are shown in Table 11B, and safety colours in Table 11C.

Table 11B: Basic identification of pipe colours

Pipe contents	Basic identification colour names (2)	BS identification colour reference BS 4800
Water	Green	12 D 45
Steam	Silver-grey	10 A 03
Oils - mineral, vegetable or animal combustible liquids	Brown	06 C 39
Gases in either gaseous or liquefied condition (except air)	Yellow ochre	08 C 35
Acids and alkalis	Violet	22 C 37
Air	Light blue	20 E 51
Other liquids	Black	00 E 53
Electrical services and ventilation ducts	Orange	06 E 51

Notes:

1.   Some colours are marginally outside the limits specified in ISO/R 508 but for practical purposes they may be used.

2.   The colour names given in column 2 are only included for guidance since different colour names may be used by different manufacturers for the same colour reference.

Table 11C: Safety colours, general

Safety colour (1)	BS colour reference BS 4800	Purpose
Red	04 E 53	fire fighting
Yellow	08 E 51	warning
Auxiliary blue	18 E 53	with basic identification colour green - fresh water (2)

Notes:

1.   The colour names given in column 1 are only included for guidance since different colour names may be used by different manufacturers for the same colour reference.

2.   Potable or non-potable.

Chapter 12　　EQUIPMENT AND APPLIANCES

12.1　The need to inspect and test

As discussed in Chapter 4, the Electricity at Work Regulations require all electrical systems to be maintained in a safe condition. The definition of a system includes all the equipment and apparatus of the fixed electrical installation of the building and equipment and appliances supplied from the fixed electrical installation. There is a wide range of portable appliance testers available to test appliances, through the expression "portable" is a little misleading as the equipment may be portable, movable, hand-held, stationary, fixed or suitable only for building-in. However, all equipment requires inspection and testing and this can generally be effected by a portable appliance tester or by the use of an insulation and continuity tester.

The Institution of Electrical Engineers publishes a Code of Practice for in-service inspection and testing of electrical equipment. This Code has been prepared for administrators with responsibility for electrical maintenance who may have little technical knowledge, and for the staff who carry out the inspection and testing. It is comprehensive guidance prepared in co-operation with a range of trade organisations and government departments including the Health and Safety Executive.

12.2　Types of test

Electrical equipment is inspected and tested at a number of stages in its life:

a)　type testing to a British Standard - carried out as part of the approval procedure for the appliance

b)　routine end-of-line testing - carried out on each appliance by the manufacturer before sale

c)　testing after repair

d)　in-service inspection and testing.

In-service inspection and testing is discussed in this chapter and is the subject of the IEE Code of Practice referred to in 12.1 above.

12.3　In-service inspection and testing

In-service inspection and testing can be categorised as follows:

a)　user checks - inspections carried out by the user. All faults found are reported and logged but no record is required if no fault is found

b)　formal visual inspections - these are inspections without tests, the results of which satisfactory or not satisfactory are recorded

c)　combined inspection and tests, the results of which are recorded.

12.4   Frequency of inspection and testing

The Electricity at Work Regulations do not specify the frequency of inspection and testing of equipment. The requirement is simply that equipment must be maintained so as to prevent danger. The user must determine how frequently equipment needs checking, inspecting and testing. Factors that would influence the frequency include:

   a)   the environment - equipment installed in a controlled environment such as an office will suffer less wear and tear and damage than equipment in an arduous environment

   b)   the care exercised by the users and likelihood of damage being reported

   c)   equipment construction - the robustness and suitability will affect the frequency.

Equipment constructions are divided into two main classes:

Class I equipment

This is equipment in which protection against electric shock does not rely solely on the basic insulation of the equipment, but also includes a means for connection of exposed-conductive-parts to a protective conductor (earth wire) in the fixed wiring of the electrical installation. In layman's terms, Class I equipment requires a connection with earth and this connection with earth must be maintained if the safety of the equipment is to be maintained. Because of this reliance upon a connection with earth, Class I equipment will generally require more frequent inspection and testing than Class II.

Class II equipment (identified by ▢)

Equipment in which protection against electric shock does not rely on basic insulation only but in which additional insulation such as supplementary insulation is provided. There is no provision for the connection of any exposed metalwork of the equipment to a protective conductor and no reliance upon protective devices in the fixed installation of the building.

Certain information technology Class II equipment may have a functional earth connection, necessary for the proper functioning of the equipment but which does not play a part in the electrical safety of the equipment.

Class III equipment

Equipment in which protection against electric shock relies on a supply of electricity from a separated extra-low voltage source such as an isolating transformer to BS 3535.

Table 12A provides guidance on suggested initial frequencies of inspection and testing of equipment. References in the notes below the Table are references to clauses in the IEE Code of Practice.

TABLE 12A INITIAL FREQUENCY OF INSPECTION AND TESTING OF EQUIPMENT

| | Type of Premises | Type of Equipment Note 1 | User Checks Note 2 | Class I | | Class II Note 4 | |
				Formal Visual Inspection Note 3	Combined Inspection and Testing Note 5	Formal Visual Inspection Note 3	Combined Inspection and Testing Note 5
	1	2	3	4	5	6	7
1	Construction sites 110 V } equipment } }	S	None	1 month	3 months	1 month	3 months
		IT	None	1 month	3 months	1 month	3 months
		M#	weekly	1 month	3 months	1 month	3 months
		P#	weekly	1 month	3 months	1 month	3 months
		H#	weekly	1 month	3 months	1 month	3 months
2	Industrial including commercial kitchens	S	weekly	None	12 months	None	12 months
		IT	weekly	None	12 months	None	12 months
		M	before use	1 month	12 months	3 months	12 months
		P	before use	1 month	6 months	3 months	6 months
		H	before use	1 month	6 months	3 months	6 months
3	Equipment used by the public	S	Note 6[+]	monthly	12 months	3 months	12 months
		IT	Note 6[+]	monthly	12 months	3 months	12 months
		M	Note 6[+]	weekly	6 months	1 month	12 months
		P	Note 6[+]	weekly	6 months	1 month	12 months
		H	Note 6[+]	weekly	6 months	1 month	12 months
4	Schools	S	weekly[+]	None	12 months	3 months	12 months
		IT	weekly[+]	None	12 months	3 months	12 months
		M	weekly[+]	4 months	6 months	1 month	12 months
		P	weekly[+]	4 months	6 months	1 month	12 months
		H	before use[+]	4 months	6 months	1 month	12 months
5	Hotels	S	None	24 months	48 months	24 months	None
		IT	None	24 months	48 months	24 months	None
		M	weekly	12 months	24 months	24 months	None
		P	weekly	12 months	24 months	24 months	None
		H	before use	6 months	12 months	6 months	None
6	Offices and shops	S	None	24 months	48 months	24 months	None
		IT	None	24 months	48 months	24 months	None
		M	weekly	12 months	24 months	24 months	None
		P	weekly	12 months	24 months	24 months	None
		H	before use	6 months	12 months	6 months	None

(1)   S   Stationary equipment
         IT   Information technology equipment
         M   Movable equipment
         P   Portable equipment
         H   Hand-held equipment

(2)   User checks are not recorded unless a fault is found.

(3)   The formal visual inspection may form part of the combined inspection and tests when they coincide, and must be recorded see 7.2b.

(4)   If class of equipment is not known, it must be tested as Class I.

(5)   The results of combined inspections and tests are recorded see 7.2c.

(6)   For some equipment such as children's rides a daily check may be necessary.

(+)   By supervisor/teacher/member of staff

110 V earthed centre-tapped supply. 230 V portable or hand-held equipment must be supplied via a 30 mA RCD and inspections and tests carried out more frequently.

The information on suggested initial frequencies given in Table 12A is more detailed and specific than HSE guidance, but is not considered to be inconsistent with it.

12.5   Review of frequency of inspection and testing

The intervals between user checks, visual inspections and combined inspection and testing must be kept under review, particularly until patterns of failure and damage are determined. Users must be encouraged to report on faults found in their more frequent checks. After some experience, it may be possible to extend the intervals between formal visual inspections and combined inspection and tests. It is suggested that there is unlikely to be any advantage in reducing the intervals between user checks.

12.6   Records

There is no specific requirement in the Electricity at Work Regulations to keep records of equipment and of inspections and tests. However, the Memorandum of Guidance on the Electricity at Work Regulations prepared by the Health and Safety Executive advises that records of maintenance including test results should be kept throughout the working life of equipment. Without such records, management cannot review the frequency of inspection and testing and determine whether the intervals need to be reduced or may be increased. It is also practically impossible to have any confidence that all equipment has been inspected and tested unless good records are kept and equipment is marked. The following records are recommended:

    a)   a register of all equipment
    b)   a record of formal inspection and tests
    c)   a repair register
    d)   a record of all faulty equipment
    e)   all equipment should be permanently and uniquely marked or labelled.

12.7   Inspecting and testing

By definition, user checks will be carried out by the user of the equipment. The formal visual inspections and combined inspection and tests may be carried out in-house if there are suitably competent staff. The IEE Code of Practice for in-service inspection and tests provides guidance on the experience and training required to carry out formal visual inspections and combined inspections and tests.

12.8   User checks

The user check is an important safety precaution. Many faults can be identified by a visual inspection. The user is the person most familiar with the equipment and may be in the best position to know if it is in a safe condition and working properly. No record is made of a user inspection unless some aspect of the inspection is unsatisfactory. The user inspection should proceed as follows:

    (a)   consider whether the user is aware of any fault in the equipment and whether it works properly

(b)   disconnect the equipment if appropriate (as described in Section 14.4 of the IEE Code of Practice)

(c)   Inspect the equipment in particular looking at:

   (i)   the flex (if fitted). Is it in good condition? Is it free from cuts, fraying and damage? Is it where it could be damaged, is it too long, too short or in any other way unsatisfactory? Does it have inadequate joints?

   (ii)   the plug (where fitted). Is the flexible cable secure in its anchorage? Is it free from any sign of overheating? Is it free from cracks or damage?

   (iii)   the socket-outlet or flex outlet. Is there any sign of overheating? Is it free from cracks and other damage?

   (iv)   the appliance - does it work? Does it switch on and off properly? Is it free from cracks, contamination damage to the case, or damage which could result in access to live parts? Can it be used safely?

   (v)   is the user satisfied that the equipment works properly?

   (vi)   is the equipment suitable for its environment?

   (vii)   is the equipment suitable for the work it is required to carry out?

(d)   Take action on faults/damage

Faulty equipment must be:

(1)   switched off and unplugged from the supply
(2)   labelled to emphasise that it must not be used (if possible, move it to a place of safe keeping)
(3)   reported to the responsible person.

Note: If equipment is found to be damaged or faulty on inspection or test, an assessment must be made by a responsible person as to the suitability of the equipment for the use/location. Frequent inspections and tests will not prevent damage occurring if the equipment is unsuitable. Replacement by suitable equipment is required.

12.9   Formal visual inspections and combined inspections and tests

Before formal visual inspections and combined inspections and tests are carried out in-house, the matter needs to be investigated thoroughly and a copy of the IEE Code of Practice for In-Service Inspection and Testing will be necessary to make an assessment.

Chapter 13   ELECTROMAGNETIC COMPATIBILITY

13.1   Introduction

Any change of electric current will cause a change in the strength of its associated electromagnetic field which in turn can induce voltages and currents in other conductors in the field. The value of the induced voltage and current is dependent upon the rate of change of current (di/dt) in the conductor causing the electromagnetic changes. Such changes in current may be caused by lightning, switching, starting, faults on the installation and, in the case of alternating current, by the sinusoidal changes in the current.

Many precautions can be taken to reduce such effects and these generally are best introduced at the design stage. As a maintenance activity, after the initial installation, it can be more difficult to reduce electromagnetic interference.

Guidance is given on reducing electromagnetic interference (EMI) in installations of buildings in IEC Standard IEC 364-4-444 Protection against Electromagnetic Interferences (EMI) in installations of buildings.

Power cables carrying large currents with high rates of change, for example the starting currents of lift motors or currents whose waveforms are chopped by solid-state devices, can induce voltages in the conductors of other equipment. These induced voltages may be significant in information technology systems and in particular in the data transmission elements of such systems.

The general rules for avoiding EMI are:

i)   purchase equipment that meets the emission standard BS EN 50081 and the immunity standard BS EN 50082

ii)   separate potential sources of interference from sensitive equipment

iii)   separate sensitive equipment such as IT communication cables from power cables subject to rapid changes of current, e.g. lift motor supplies

iv)   fit filters and surge protection in sensitive equipment power supply circuits

v)   avoid unwanted tripping of circuits by careful selection of device characteristics including, if necessary, time delays

vi)   bond metal enclosures (but see viii) and x)) and screen sensitive equipment

vii)   separate power and signal cables and always cross over at right angles - see Table 13A

viii)   avoid inductive loops

ix)   use screened and/or twisted pair data cables

x)   make bonding connections as short as possible

xi)   enclose single core conductors in earthed metal enclosures or equipment. Note, this cannot be done for power cables without taking special precautions

xii)   do not use TN-C systems (see Part 2 of BS 7671 for definition)

xiii) attempt to arrange for all metal service pipes and cables to enter the building at the same place

xiv)   bond screens, metal pipes etc.

xv)   for communications between systems with independent earthing arrangements, use fibre-optic or non-conducting links.

The basic principle of avoiding the emission of interference and making cables less susceptible to emitted signals (immunity) is to keep all the conductors of a particular system or circuit in very close proximity - phase conductors, neutral conductors, functional earth conductors and protective earth conductors.

This is the principle of the twisted pair. Here, the current in one conductor is matched by the return current in the other conductor. Similarly, changes in the electromagnetic field caused by other equipment induces equal and opposite voltages in each conductor of the pair, cancelling out the effect at the equipment.

In practice, there could be a number of phase conductors and both protective and functional conductors and there may be difficulties in keeping them all in close proximity. This is particularly true of protective conductors, where there may be parallel paths and loops.

The use of a clean earth can be beneficial, i.e. a protective conductor connecting sensitive equipment directly to the main earth. Functional earth conductors should be run in proximity with the phase and neutral conductors supplying the equipment for the reasons explained above. They can be particularly beneficial if the protective conductor is liable to pick up induced voltages from other equipment. Loops in protective conductors will naturally act as aerials picking up radiated field changes. Figure 13.1 illustrates measures that could be undertaken to protect against EMI in an existing building.

Figure 13.1 Illustration of measures for an existing building

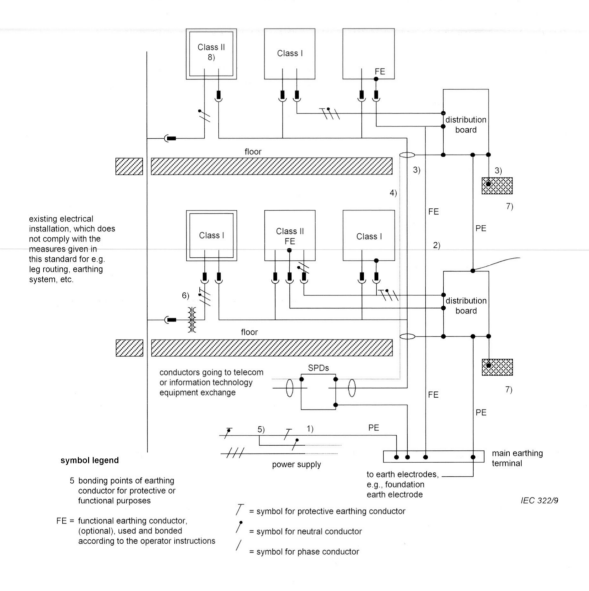

Description of the illustrated measures		Reference
Subclause		
444.3.13	Cables and metal pipes enter the building at the same place	1)
444.3.8	Common route with adequate separations and avoidance of loops	2)
IEC 1000-2-5; 444.3.10	Bondings as short as possible, and use of earthed conductor parallel to a cable	3)
444.3.9	Signal cables screened and/or conductors twisted pairs	4)
444.4.12	Avoidance of TN-C beyond the incoming supply point	5)
444.4.3	Use of transformers with separate windings	6)
IEC 364-5-548, annex B	Local horizontal bonding system, if available	7)
444.4.2	Use of class II equipment	8)

The Institution of Electrical Engineers has published a report "Electromagnetic Interference, Report of the Public Affairs Study Group" and the separations given in Table 13A are taken from this. This report advises that power and signal cables should not be grouped together in the same conduit or on the tray, and where of necessity parallel cable runs exist between power and signal cables, the separation should be not less than the minimum values given in Table 13A. It should be noted that the effects of interference will not only depend upon the field strength changes caused by the power cables, but also to the sensitivity of the information technology equipment. All such equipment should meet the immunity levels of BS EN 50082.

Table 13A   Separation between power and signal cables

Power cable Voltage	Minimum separation between power and signal cable (metres)	Power cable current amperes A	Minimum separation between power and signal cable (metres)
115 V	0.25	5	0.24
240 V	0.45	15	0.35
415 V	0.58	50	0.5
3.3 kV	1.1	100	0.6
6.6 kV	1.25	300	0.85
11.0 kV	1.4	600	1.05

13.2   Interference with telecommunications and data processing equipment

Guidance on the separation of power cables from telecommunication cables is given in BS 6701: 1994 Code of Practice for the Installation of Apparatus Intended for Connection to Telecommunication Systems. This recommends minimum separation distances, as reproduced here in Table 13B.

Table 13B   Separation distances between power and telecommunication circuits (from BS 6701)

External cables

Minimum separation distances between external low voltage electricity supply cables operating in excess of 50 V a.c. or 120 V d.c. to earth, but not exceeding 600 V a.c. or 900 V d.c. to earth, and Telecommunications cables.

Voltage to earth	Normal separation distances	Exceptions to normal separation distances, plus conditions to exception
Exceeding 50 V a.c. or 120 V d.c., but not exceeding 600 V a.c. or 900 V d.c.	50 mm	Below this figure a non-conducting divider should be inserted between the cables.

Internal cables

Minimum separation distances between internal low voltage electricity supply cables operating in excess of 50 V a.c. or 120 V d.c. to earth, but not exceeding 600 V a.c. or 900 V d.c. to earth and Telecommunications cables.

Voltage to earth	Normal separation distances	Exceptions to normal separation distances, plus conditions to exception
Exceeding 50 V a.c. or 120 V d.c., but not exceeding 600 V a.c. or 900 V d.c.	50 mm	50 mm separation need not be maintained, provided that (i) the LV cables are enclosed in separate conduit which if metallic is earthed in accordance with BS 7671 **OR** (ii) the LV cables are enclosed in separate trunking which if metallic is earthed in accordance with BS 7671 **OR** (iii) the LV cable is of the mineral insulated type or is of earthed armoured construction.

Notes:

1.   Where the LV cables share the same tray then the normal separation should be met.

2.   Where LV and telecommunications cables are obliged to cross, additional insulation should be provided at the crossing point; this is not necessary if either cable is armoured.

112

Chapter 14   LIGHTNING PROTECTION INSTALLATIONS

14.1   Design and installation

British Standard 6651: 1992 Code of Practice for Protection of Structures against Lightning provides comprehensive advice on the protection of structures against lightning, including :

estimation of the need for protection
system design
inspection and testing
records.

This Publication considers the periodic inspection and testing and records. The advice provided in this Chapter is based on that given in BS 6651.

14.2   Records

The following records should be available to any maintenance engineer responsible for the lightning protection installation :

1.   Drawings of the installation*
2.   Information on the soil, in particular the presumed soil resistivity (Ω m)
3.   The type and position of the earth electrodes
4.   The results of all previous tests, including environmental conditions at the time of testing
5.   Details of any repairs carried out.

*The scale drawings should be up-to-date and show any additions or changes made.

14.3   Maintenance and upkeep

Periodic inspection and tests should be carried out at fixed intervals not exceeding 12 months. Inspection should also be carried out following a known lightning strike to the installation.

Inspection

The inspection should confirm that the installation is installed as per the record drawings and is complete in all respects, with particular attention being paid to the continuity of the protective conductors.

Particularly, the inspector should check that all earthing connections remain in good condition and should identify in his report, any evidence of corrosion or conditions likely to lead to deterioration.

All changes, alterations or additions to the building structure which may require changes to the lightning protection system, including change of use, particularly such changes as the erection of masts, aerials or chimneys must be reported.

Testing

Testing is an integral part of maintenance which is carried out to confirm the visual inspection to check that there is continuity, and that the resistance to earth is within the limits required by the British Standard. The results should be compared with the previous test results and any serious changes, increases or decreases, investigated. Environmental conditions should be recorded when carrying out tests. If it is thought that the previous tests were carried out in favourable conditions and that the resistance would be higher say in the height of summer when the ground is dry, then arrangements should be made to carry out measurements under the worst conditions to be expected.

It is a requirement of BS 6651 (Clause 16.1) that an earth electrode should be connected to each down conductor of the installation. The resistance to earth of each of these earth electrodes should not exceed the resistance value given below.

Maximum value of electrode resistance ≤ 10 x No. of earth electrodes.

For example, if there are 10 electrodes in the installation, the resistance of each individual electrode should not exceed 100 ohms and if there are 20 electrodes in an installation, the resistance of each individual electrode should not exceed 200 ohms.

The resistance to earth of the complete installation, comprising all the electrodes, should not exceed 10 ohms. This value must be achieved when not bonded to other services.

It is emphasised in the Standard that, before disconnecting the lightning protection earth, it should be tested using a sensitive voltage testing device to ensure that it is not live.

If the original drawings of the installation are not available, it is necessary to take specialist advice as to compliance with BS 6651, as it may not be immediately apparent how the protection is provided. For example, in a steel frame structure the steel frame members may be acting as down conductors and their incorporation in the foundations of the building might lead to unacceptably high resistance without additional earth electrodes.

It is important to reduce the resistance to earth to 10 ohms or less as this reduces potential gradients around the earth electrodes when lightning currents are discharged. It may also reduce the risk of side flashing.

14.4   Bonding to other services

The lightning protection system is required to be bonded to the metalwork if any, of incoming services. A typical installation is shown in Figure 14A.

Figure 14A - Bonding to services

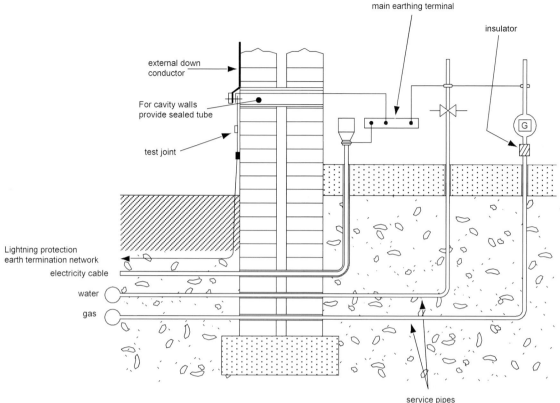

From Figure 30. Diagram showing bonding to services (gas, water and electricity) of BS 6551

IEC 364-4-4-43 advises that where an installation is supplied by a complete low voltage underground system or does not include overhead lines, no additional protection against overvoltage of atmospheric origin is necessary.

Where an installation is supplied by or includes a low voltage overhead line, additional protection is required if there are more than 25 thunderstorm days per year. BS 6651 provides figures of thunderstorm days for the UK of between 5 and 20 and, consequently, no additional protection is normally required in the UK even when the installation is supplied by a low voltage overhead line.

Should it still be wished to provide protection against overvoltage, then this should be installed close to the origin of the installation, either in the overhead line or in the building installation.

Chapter 15 LIGHTING MAINTENANCE

15.1 Introduction

A lighting installation should be maintained to keep its visual performance within the design limits. The designer will have selected a certain illumination level for the particular activity and presumed a frequency of lamp replacement and a frequency of cleaning. The frequency of lamp replacement and cleaning will be appropriate to the environment including accessibility and the type of luminaires (light fittings). When assessing maintenance requirements the first step is to seek information on the initial design assumptions. Maintaining a lighting installation as intended will ensure the efficiency of the installation is not degraded. Reduced maintenance may result in reduced operative performance and, in extremes, lead to danger. Maintaining a lighting installation in good order is also important for maintenance of staff morale and in providing a good impression to customers. Flickering, failed and discoloured lamps may discourage staff and turn customers away.

There are two aspects to luminaire maintenance:

1.   luminaire cleaning
2.   lamp replacement.

15.2 Luminaire cleaning

Frequency

The frequency of luminaire cleaning depends upon three factors:

1.   The type of luminaire and its inherent ability to maintain light output over a period for a given environment
2.   The environment (whether it is clean, normal or dirty)
3.   The permitted (or assumed for design purposes) reduction in light output before luminaire cleaning is required.

The reduction in light output as a proportion of initial light output is called the luminaire maintenance factor (LMF) and is given in Table 15A.

The following procedure can be followed for determining the intervals between luminaire cleaning :-

1)   Determine luminaire maintenance category from Table 15B.
2)   Assess the environment, see Table 15C.
3)   Identify the luminaire maintenance factor.

Reference to Table 15A will then give the permissible period between luminaire cleaning.

The designer of the lighting installation will have selected a luminaire maintenance factor (LMF), however, in the absence of such information a factor of 0.8 is usually assumed.

116

Table 15A   Luminaire maintenance factor (LMF)
(Light output as a proportion of initial light output - see note 1)

Elapsed time Between Cleanings in Years	0.5			1.0			1.5			2.0			2.5			3.0		
Environment Note 2	C	N	D	C	N	D	C	N	D	C	N	D	C	N	D	C	N	D
Luminaire category																		
A	.95	.92	.88	.93	.89	.83	.91	.87	.80	.89	.84	.78	.87	.82	.75	.85	.79	.73
B	.95	.91	.89	.90	.86	.83	.87	.83	.79	.84	.80	.75	.82	.76	.71	.79	.74	.68
C	.93	.89	.83	.89	.81	.72	.84	.74	.64	.80	.69	.59	.77	.64	.54	.74	.61	.52
D	.92	.87	.83	.88	.82	.77	.85	.79	.73	.83	.77	.71	.81	.75	.68	.79	.73	.65
E	.96	.93	.91	.94	.90	.86	.92	.88	.83	.91	.86	.81	.90	.85	.80	.90	.84	.79
F	.92	.89	.85	.86	.81	.74	.81	.73	.65	.77	.66	.57	.73	.60	.51	.70	.55	.45

Notes:

1.   Luminaire maintenance factor (LMF) is given by :

$$LMF = \frac{\text{light output after a specified time}}{\text{initial light output}}$$

2.   Environment references:

C = Clean environment
N = Normal environment
D = Dirty environment

Table 15B Luminaire maintenance categories

Category	Representation	Description
A		The light source is in free air. There are no reflector surfaces, diffusers, or covers to be contaminated.
A		GLS lamp
B		White painted metal reflectors with slots in the top to create air currents to keep the reflectors clean.
B		High bay luminaires
B		PAR 38 reflector lamp
B		GLS lamp with reflector
B		Compact fluorescent with reflector

Table 15B   Luminaire maintenance categories continued

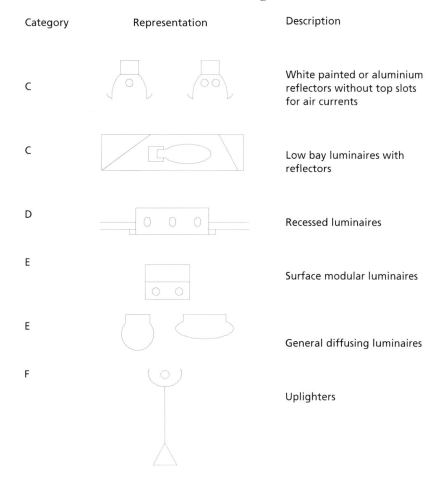

Category	Representation	Description
C		White painted or aluminium reflectors without top slots for air currents
C		Low bay luminaires with reflectors
D		Recessed luminaires
E		Surface modular luminaires
E		General diffusing luminaires
F		Uplighters

Table 15C   The Environment

Category	Environment	Examples
C	clean	offices, shops, laboratories
N	normal	light industrial, outdoor
D	dirty	contamination by smoke, dust etc. e.g. foundries, rubber processing

Assuming a luminaire maintenance factor of 0.8, cleaning intervals have been determined in Table 15D for selected luminaires and maintenance types.

119

Table 15D Cleaning intervals

		Cleaning Intervals - Years, LMF = 0.8		
Luminaire Maintenance Category	Typical Luminaire type	Environment		
		Clean C	Normal N	Dirty D
A	fluorescent no reflector	3	2	1
B	High bay	3	2	1
C	fluorescent aluminium reflector	2	1	½
D	Recessed	2	1	½
E	General	3	3	2
F	Uplighter	1	1	½

Wall and ceiling cleaning

Room lighting levels, as well as depending upon the cleanliness of the luminaires also depend on the cleanliness of the room, particularly ceilings and walls. This factor is called the room surface maintenance factor (RSMF). The designer will have assumed a factor for this and may well have assumed a cleaning time for the walls and ceiling. Again, the intervals between the cleaning of floors and ceilings will depend upon the environment and to a lesser degree the nature of the lighting, i.e. whether it is direct or indirect, and on the size of the room. Indirect lighting reflecting from a ceiling is very dependent upon the cleanliness of the ceiling or the surface from which the lighting is being reflected, and dirty walls will have a lesser impact on a large room than on a small room.

Precautions when cleaning luminaires

Extreme caution should be exercised when cleaning all surfaces of luminaires. Some surfaces are very susceptible to abrasion, for example, polished (unanodised) aluminium is easily scratched, as are some plastics.

The maintainer should experiment on a small test area with the proposed cleaning method before starting.

Care is required in handling plastics, as they tend to become brittle with age. Depending on the environment or light source, some plastics may also turn yellow. There is no successful way of cleaning when this happens and replacement must be considered.

Cleaning methods

Aluminium reflectors should be washed with a warm, soapy solution and rinsed thoroughly before being air dried. Plastic opal or prismatic lenses should be cleaned with a damp cloth (using non-ionic detergent and water) and treated with antistatic polish or spray and allowed to dry. Vitreous enamel, stove enamel and glass optics should be wiped with a damp cloth using a light concentration of detergent in water.

For best results, louvre (rectangular and square cell) optics should be cleaned by removing the louvre and dipping it in a warm water solution.

Specular finished (particularly plastic) louvres are very difficult to clean, and appearance deteriorates over the years. Therefore, they should be used only where air quality is very clean, such as new office buildings, banks, etc.

Cleaning agents

Choice of cleaning materials and methods is determined by the type of dirt to be removed and the type of material to be cleaned. For plastic materials a final treatment with antistatic substance is recommended.

General Cleaning - The first and most commonly used is a dry chemical detergent with additives in different concentration levels. It is an advantage to use compounds that require no rinsing after the wash.

Heavy Duty Cleaning of Oil Concentrations (e.g. in auto garages, oily factories, etc.) - The second type of cleaner is a heavy duty liquid cleaner which may contain detergents, solvents and abrasives. It is particularly useful for the removal of oily dirt, but must be tested to ensure that it does not damage materials or leave deposits.

Excessively Oily Industrial Conditions - In some heavy oily applications, the use of a high pressure steam cleaner is practicable provided the scheme has been designed with this cleaning technique in mind.

15.3   Lamp replacement

General

Three factors are particularly important when considering the frequency of lamp replacement. These are:

1. The lamp survival factor, LSF (the proportion of lamps still working after a specified burning time).
2. The lamp lumen maintenance factor, LLMF (the proportion of the initial light output being maintained after a specified burning time due to deterioration (ageing) of the lamp).
3. The cleaning frequency.

Lamp survival factor (LSF)

Lamps fail in service, some more frequently than others. This is particularly important with incandescent lamps, as can be seen from Table 15E. For other types of lamp failure is unlikely until the burning time has exceeded some thousands of hours. What is likely to precipitate lamp change is the reduction in the initial light output and the environmental effects of failed or flickering lamps. Whilst a failed or flickering lamp may not make much difference to the illumination level it is both disturbing and distracting to staff and customers.

The maintenance engineer has to decide what is an acceptable lamp survival factor not only from a lumen output point of view but also from appearance. One in twenty lamp failures may be just acceptable in a light industrial workshop but would probably be unacceptable in an office environment. The longer the intervals between group lamp replacement the greater are the costs associated with individual lamp replacements. The maintenance engineer cannot generally refuse a request to replace a failed lamp in an office environment.

Lamp lumen maintenance factor (LLMF)

The lamp lumen maintenance factor (LLMF) can be confused with the luminaire maintenance factor (LMF). The LLMF is concerned with reductions in light output of the lamp (not luminaire) as a result of ageing. Cleaning does not change the LLMF. The luminaire maintenance factor (LMF) is concerned with dirt on the luminaire.

As a lamp burns, it discolours and its light output reduces. The LLMF is a measure of this and is given by:

$$LLMF = \frac{\text{light output after a specified burning time}}{\text{initial light output}}$$

Table 15E Lamp lumen maintenance factors and lamp survival factors

Lamp Type	Factors note	Burning (thousand hours)															
		0.1	0.5	1.0	1.5	2.0	4.0	6.0	8.0	10.0	12.0	14.0	16.0	18.0	20.0	22.0	24.0
Incandescent GLS	LLMF	1.00	.97	.93	.89												
	LSF	1.00	.98	.50	.30												
Fluorescent Multi and Tri-phosphor	LLMF	1.00	.98	.96	.95	.94	.91	.87	.86	.85	.84	.83	.81				
	LSF	1.00	1.00	1.00	1.00	1.00	1.00	.99	.95	.85	.75	.64	.50				
Fluorescent Halophosphate	LLMF	1.00	.97	.94	.91	.89	.83	.80	.78	.76	.74	.72	.70				
	LSF	1.00	1.00	1.00	1.00	1.00	1.00	.99	.95	.85	.75	.64	.50				
Mercury	LLMF	1.00	.99	.97	.95	.93	.87	.80	.76	.72	.68	.64	.61	.58	.55	.53	.52
	LSF	1.00	1.00	1.00	1.00	.99	.98	.97	.95	.92	.88	.84	.80	.75	.68	.59	.50
Metal Halide	LLMF	1.00	.96	.93	.90	.87	.78	.72	.69	.66	.63	.60	.56	.52			
	LSF	1.00	1.00	.97	.96	.95	.93	.91	.87	.83	.77	.70	.60	.50			
High Pressure Sodium	LLMF	1.00	.98	.98	.97	.96	.93	.91	.89	.88	.87	.86	.85	.83	.82	.81	.80
	LSF	1.00	1.00	1.00	1.00	.99	.98	.96	.94	.92	.89	.85	.80	.75	.69	.60	.50
High Pressure Sodium-Improved Colour	LLMF	1.00	.99	.97	.95	.94	.89	.84	.81	.79	.78						
	LSF	1.00	1.00	1.00	.99	.98	.96	.90	.79	.65	.50						

Note:
LLMF -   (lamp lumen maintenance factor) : proportion of the initial light output emitted after a specified burning time
LSF   -   (lamp survival factor) : proportion of lamps surviving after a specified burning time.

123

Cleaning frequency

It is cost effective to replace lamps when carrying out routine cleaning of the luminaires. As a result, it is almost always sensible to arrange lamp replacement during routine cleaning, but perhaps not at every routine cleaning.

Example calculation of lamp replacement

Having decided upon:

> the lamp survival factor (LSF)
> the lamp lumen maintenance factor (LLMF)

the frequency of lamp replacement can be calculated.

Consider a fluorescent lamp installation (multi phosphor) and assume the LSF must be 0.95 or greater and the LLMF 0.8. From Table 15E:

1)   LSF 0.95: lamps must be replaced before 8000 hrs.
2)   LLMF 0.8: lamps must be replaced before 16000 hrs.

Therefore replace at lowest of 1 and 2, that is 8000 hrs.

The lamp burning hours need to be estimated. In most commercial premises little natural light is available and hours of occupancy are a very good guide to hours of burning.

Assuming a 9 hr day, 6 days a week for 50 weeks

> 9 x 6 x 50 = 2700 hrs burning per annum.

In the example above, the lamps would need replacing after

$$\frac{8000}{2700} = \text{i.e. 3 years}$$

After an estimation of the cleaning intervals as described in section 15.2 decisions can be made on the frequency of the combined activities of cleaning and lamp replacement. In Table 15F examples have been prepared that include the calculation of cleaning intervals, and lamp replacement frequency.

Table 15F   Examples of selection of lamp change and luminaire cleaning frequencies.

Location	Luminaire Type	Occupancy Description	Days/year	Hours/day	Lamp burning time hrs/year	Adopted Lamp Survival Factor LSF	Burning hours to LSF from table 15E	Years to LSF	Adopted Lamp Lumen Maintenance Factor LLMF	Burning hours to LLMF from table 15E	Years to LLMF	Selected Luminaire Maintenance Factor	Environment	Time between cleaning from table 15A	Selection Lamp change	clean
Office	Fluorescent multi phosphor B	6 days per week	300	9	2700	0.95	8000	2.96	0.80	16000	5.9	0.8	Clean	3.0	3	3
Light Industry	Fluorescent muli phosphor B	2 shifts, 6 days per week	300	16	4,800	0.9	9000	1.88	0.85	10,000	2.0	0.8	Normal	2.0	2	2
Street lighting	Sodium E	All night half night	365		4000 2000	0.9 0.9	10000 10000	2.5 5	0.80 0.80	24000 24000	6 12	0.8 0.8	Normal Normal	3+ 3+	3(1) 6(1)	3 3

1) with failed lamps replaced as necessary

125

15.4   Lamp disposal

Chapter 16 discusses lamps as special waste. In this section practical approaches to lamp disposal are described.

Accidental breakage of a lamp

If a lamp is broken, only normal good housekeeping to prevent injury and perhaps product contamination from broken glass is required. The usual precautions need to be taken when disposing of the broken glass.

Disposal of discharge lamps

Where individual lamps are removed from service, the principal hazard is the risk of breakage of glass. This can be reduced by placing the old lamps in the packaging provided with the new lamps. Individual lamps can be disposed of with normal waste. Where there is bulk disposal of lamps, they are normally disposed of by fragmentation, followed by dispersal of the released chemicals with water and discharge of the effluent and debris after any necessary treatment. The guidance on disposal of lamps given here applies to the lamps typically available for domestic, commercial and industrial use, such as fluorescent lamps, high pressure mercury e.g. MB, MBF, MBR and MBTF, metal halide, high pressure sodium e.g. SON, and low pressure sodium e.g. SOX and SL9 and SLI.

The hazards

The breaking of lamps gives rise to hazards associated with :

1.   Flying glass

2.   Fire and combustive explosion

3.   Toxic and corrosive substances.

There is an ongoing hazard from flying glass and the handling of glass fragments, which is the major danger throughout the disposal operation.

Low pressure sodium lamps (SOX) contain elemental sodium and when these lamps are broken, it is necessary to add water in excess to ensure that all sodium metal is oxidised. The reaction produces hydrogen and sodium hydroxide solution in which rapid ignition of the hydrogen is likely to occur. If this does not happen, a flammable gas/air mixture could accumulate and an explosion hazard arise. In view of the relatively small quantities of sodium contained in discharge lamps (see Table 16A), the risk of fire or explosion is low providing the precautions recommended below are implemented.

Note: High pressure sodium lamps also contain sodium although less than SOX and the sodium is contained in the inner quartz tube, which is difficult to break. Consequently it is recommended that high pressure sodium lamps are not fragmented.

Discharge lamps contain small quantities of toxic materials, including lead and mercury which may be released as dust or vapour. Sodium hydroxide spray may be emitted during the disposal of sodium lamps and precautions must be taken to avoid contact of the sodium hydroxide with skin, eyes etc.

Proper facilities need to be provided for the disposal of lamps using only trained staff and professionally constructed equipment. Certain local authorities make arrangements for fragmentation disposal of lamps and this may be the preferred option for an employer.

15.5   Precautions for the disposal of lamps

The following precautions should be taken when lamps are being fragmented prior to disposal. Additional precautions are required for sodium lamps.

1.   The fragmentation and disposal should only be carried out by trained staff with written procedures and the provision of appropriate equipment and protective clothing.

2.   The fragmentation of lamps should be carried out only in specially constructed containers, well ventilated in the open air, where contaminants emitted can disperse safely, for example, without entering buildings or crossing into public areas.

Adequate protection should be provided against flying glass and other fragments, for example, by the design of the container or the provision of screens.

Persons not trained or not involved in the disposal operation should not be allowed to enter the area.

The operator and other persons should be provided with appropriate protective clothing including protection for eyes, face, neck, hands and arms. So far as is reasonably practicable, open fragmentation containers should not be approached until the debris has been effectively rinsed with water. Where the water is added manually then application by hose from a distance is recommended.

Respiratory protection should be provided if there is a foreseeable toxic risk.

Note: The inner tube of a mercury lamp contains mercury. The inner tube is difficult to break and consequently is not normally broken in the fragmentation process. No special effort should be made to break this inner container to minimise the risk of release of the contents.

Additional precautions for sodium lamps

Low pressure sodium lamps contain sodium which reacts with water evolving heat. These lamps should be broken before disposal. The arc tube of high pressure sodium lamps is tough and should not be broken manually before disposal.

The following precautions, in addition to those above, should be taken to minimise the risk of fire and explosion.

Work in a dry atmosphere. No more than 20 lamps should be carefully broken into a large dry container. When the container is approximately one quarter full of lamp debris, the operator should fill it with water from a safe distance, for example with a hose. The water will react with the sodium. The water may be disposed of as a weak caustic soda solution and the debris disposed of as glass.

The container for fragmentation should be of a non-combustible material, adequately ventilated particularly at high level, e.g. open topped and located in a safe, well ventilated position in the open air to ensure that the hydrogen and any contaminants are safely dispersed.

The container used for fragmentation should be specifically designated for this purpose. It should not contain any combustible material or other substances which might react dangerously with the materials released in the operation.

Sufficient water should be added to ensure that all the sodium has been completely reacted.

Personnel should not approach the container until the reaction with the water has been completed.

Sources of ignition should be removed from the vicinity. Smoking and naked lights should be prohibited and suitable notices should be posted to this effect.

If the resultant alkaline solution is to be neutralised by the addition of dilute acid, a proper system of work should be prepared and staff properly trained.

Disposal of fragmented waste

It is recommended that arrangements are always made to dispose of fragmentation waste, in co-operation with the local authority, at appropriately licensed sites. The relevant waste disposal authority should be prepared to provide guidance.

The local drainage authority must be advised of any proposals to dispose of effluent to ensure that proper control measures, for example the neutralisation of alkaline solutions, are effected and permitted discharge levels are not exceeded.

References

1.   Disposal of discharge lamps, HSE information document HSE 253/3.

2.   LIF Technical Statement No. 10:
      CO SHH Regulations 1988 Health and Safety at Work etc. Act 1974 and disposal of lamps.

Chapter 16   SPECIAL WASTE

16.1   Special waste regulations 1996

The Special Waste Regulations 1996 implement European Council Hazardous Waste Directive 91/689/EEC on hazardous waste. Waste becomes special waste and subject to the controls if it is any of the following:

1)   on the hazardous waste list - attached to Schedule 2 of the Regulations
2)   it contains a listed substance - Schedule 3 of the Regulations
3)   it has certain characteristics - see Schedule 4 of the Regulations.

The special characteristics could be a concentration of a corrosive substance, a flashpoint below a specified temperature, a concentration of irritant, a concentration of carcinogenic substances, etc.

The Special Waste Regulations require that a consignment note system be implemented to ensure that all special wastes are safely managed from "cradle to grave". This publication does not attempt to deal with wastes other than those routinely encountered during electrical maintenance.

16.2   Disposal of discharge lamps

Fluorescent and discharge lamps contain elements of environmental interest as shown by Table 16A below.

Table 16A - Elements of environmental interest contained in fluorescent and discharge lamps

Element	Quantity per lamp - milligrammes					
	Fluorescent		Mercury High Pressure	Metal Halide	Sodium High Pressure	Sodium Low Pressure
	Halo Phosphate	Tri-Phosphor				
Mercury	35	35	20	30	20	-
Sodium	-	-	-	3	5	400
Lead	Note 4	Note 4	Note 4	Note 4	Note 4	Note 4
Antimony	30	20	-	-	-	-
Barium	5	20	2	15	20	50
Indium	3	3	-	-	-	30
Manganese	60	30	-	-	-	-
Strontium	30	30	50	50	-	10
Thallium	-	-	-	10	-	-
Thorium	-	-	1	1	1	-
Vanadium	-	-	100	50	-	-
Yttrium	-	700	170	240	2	-
Rare Earths	-	200	20	30	-	-

Notes:

1. Rounded values have been used, based on the average for each lamp type. There will be some variations between ratings and between manufacturers.

2. Values relate to 1987 technology, but quantities per lamp are not likely to increase.

3. The Table does not include constituents of glass/quartz envelopes.

4. For caps with solder, add approximately 1 gramme of lead per lamp.

5. Mercury, sodium and lead are present as metals; all others are constituents of inorganic compounds.

It is clear that care has to be taken with respect to the disposal of lamps and if in doubt all should be treated as special waste. The disposal process normally involves the breaking of lamps and there are hazards associated with this. The disposal of lamps is outlined in Section 15.4.

Fluorescent lamps

If a lamp is accidentally broken or an individual lamp replaced the lamp can be disposed of as the normal waste. Clearly, good housekeeping is required to prevent injury from broken glass etc. Where there is a bulk lamp change they should normally be disposed of by fragmentation followed by release of dispersal of released chemicals with water and discharge of the effluent and debris after the treatment. The local waste authority should be consulted with respect to the disposal of the lamps. Waste fluorescent lamps are not special waste as such and consequently it is normally accepted that small numbers of fluorescent lamps are disposed of with general waste at any appropriate local licensed site. However, larger consignments may well be regarded as hazardous waste and should be disposed of at a licensed site and the advice of the waste disposal authority sought.

The dissolved chemicals and effluent from the fragmentation of lamps should be disposed of in accordance with the requirements of the local drainage authority. Control measures may be needed to be adopted to ensure that permitted discharge levels of the effluent are not exceeded.

The Environment Agency has issued a technical assessment of waste Number TEMP27 for mercury fluorescent tubes. The summary advises:

Fluorescent tubes

Fluorescent tubes that do not contain sodium are not special waste. (Note tubes and lamps which contain sodium are the subject of a separate assessment to which reference should be made.) Mercury may however be released as a vapour when tubes are broken and although not special waste, attention is drawn to the general guidance on tube management and disposal contained in

HSE Guidance "Disposal of Lamps" HSE 253/3 and obligations under the duty of care. The Guidance given in HSE 253/3 is generally reproduced in Section 15.4.

Sodium lamps

Reference to Table 16A indicates that, particularly with respect to low pressure lamps, there is a considerable sodium content in sodium lamps. Sodium lamps are likely to be special waste not only for their content of listed substance but also because the sodium when exposed to air will cause fire, burns etc. All sodium lamps should be treated as special waste and disposed of by crushing and water as described in Section 15.4 and the residue disposed of by agreement with the local authority.

16.3   Capacitors

Generally, capacitors manufactured until about 1976 contained polychlorinated biphenyls (PCBs). Capacitors, including those installed in power factor correction equipment and in fluorescent and discharge luminaires, do not carry a label identifying the PCB content. They will typically be date coded. Government regulations banned the sale of PCB-filled capacitors from 1 June 1986. Capacitors containing PCBs can continue to be used if the capacitor case remains serviceable. However, capacitors eventually fail and in certain circumstances PCBs may leak out. It is generally advised that all PCB filled capacitors should be replaced. Capacitors containing PCBs or PCTs are scheduled in the hazardous waste list with hazard classification N, as they are very toxic to aquatic organisms and may cause long-term adverse effects in the aquatic environment. It is necessary to contact the waste authority to obtain practical guidance on disposal of capacitors. The procedure for bulk disposal of PCBs normally involves incineration at a special plant.

If a PCB leak occurs it usually shows itself in the early stages as a brown stain. The procedure should be to isolate the luminaire from the supply and the operative should be issued with single-use polythene gloves. The gloves must be robust to give substantial protection. Rubber gloves should not be used. Operatives must wear the gloves provided and avoid direct contact with the liquid. The capacitor should be removed from the luminaire, wrapped in absorbent material, then sealed in a plastic bag together with the gloves any material used to clean the luminaire. Individual capacitors so replaced may be disposed of with normal waste unless local requirements specify otherwise.

Since all the luminaires in an installation were probably installed at the same time, it is likely that other capacitors will leak. Clearly, in these circumstances an urgent review of the lighting installation would be required and it would probably be appropriate to change the luminaires for new energy efficient fittings to provide a safe and long-term solution.

16.4 Other equipment

Transformers

Transformers may also contain PCBs or PCTs and they should generally be treated in the same way as described above for capacitors.

Oil-filled transformers

Oil wastes are scheduled hazardous waste and must be disposed of in co-operation with the local waste authority in accordance with the requirements of the hazardous waste directive and the Special Waste Regulations.

Asbestos

Many electrical products contain asbestos and it is commonly found as thermal insulation and fire barriers in buildings. Environment Agency technical assessment of waste PTEMP 04 advises that all forms of asbestos, regardless of the chemical form, are listed as Category I carcinogens in the approved list. All forms of asbestos are regarded as special waste where the asbestos content is greater than the carcinogen threshold concentration of 0.1%W/W.

References

Health and Safety Executive Information Document HSE 253/3 - Disposal of Discharge Lamps
Environment Agency Technical Assessments as follows:
TEMP27-Mercury Fluorescent Tubes
PTEMP 04 - Asbestos
PTEMP 05 - Waste containing PCBs or PCTs

Lighting Industry Federation Technical Statements Number:

1.   Capacitors containing polychlorinated biphenyls (PCBs) (fluorescent and discharge lighting)
10.   COSHH Regulations 1998 Health and Safety at Work etc. Act 1974 and disposal of lamps.

The Special Waste Regulations 1996.

Chapter 17   LEGIONELLOSIS

17.1   Publications

The guidance given in this Chapter is based on Health and Safety Commission Publication "The prevention or control of Legionellosis (including Legionnaire's Disease) reference L8(REV)". Detailed technical advice on assessing and minimising the risk from exposure to Legionellae is given in Health and Safety Executive Guidance HS(G)70 "The control of Legionellosis including Legionnaire's Disease".

17.2   Legionellosis

Legionellosis is the term used for infections caused by Legionellae pneumophila and other bacteria in the family legionellaceae. Legionnaire's disease is a pneumonia that affects persons made susceptible due to age, illness, immunosuppression, smoking etc. and it can be fatal.

Infection is considered to occur by the inhaling of Legionellae, either those in water droplets or by inhaling droplet nuclei, that is the particles left after the water from a droplet has evaporated. Most cases, in outbreaks of Legionellosis, have been attributed to infections in water services in buildings, in the water of cooling towers and water in whirlpool spas.

17.3   Legislation

The Health and Safety at Work etc. Act 1974 and the Control of Substances Hazardous to Health Regulations 1994 have requirements for preventing or minimising the risk from exposure to Legionellae. These include requirements for :

Identification and assessment of risk

Preventing or minimising the risk from exposure

Management, selection and training of personnel

The need for record keeping, and responsibilities on designers, manufacturers, importers, suppliers and installers.

Water systems creating risk

The legislation applies to any work activity or premises connected with a trade, business or undertaking where water is used or stored and where there are means of creating water droplets which may be inhaled, thereby causing a reasonably foreseeable risk of Legionellosis. Effective precautions to stop or contain transmission of droplets over an area where they might be inhaled would prevent risk. Experience has shown that the following present a risk of Legionellosis :

Water systems incorporating a cooling tower

Water systems incorporating an evaporative condenser

Hot water services

The hot and cold water services, irrespective of size, in premises where occupants are particularly susceptible e.g. old people's homes, nursing homes and general health care premises.

Humidifiers and air washes which create a spray of water droplets and in which the water temperature is likely to exceed 20°C.

Spa baths and whirlpools in which warm water is deliberately agitated and recirculated.

General plant and systems containing water likely to exceed 20°C and which may release a spray or aerosol during operation or when being maintained.

Shower heads.

If any employer or person responsible for a trade, business, or similar undertaking has water systems as described above, arrangements must be made for an assessment to be carried out to identify the risk to health and the measures that can be taken to prevent or control the risk of exposure to Legionellae.

Identification and assessment of risk

For premises with systems as described above, a suitable and sufficient assessment should be carried out to identify and assess the risk of Legionellosis from work activities and water sources. The purpose of the assessment is to enable valid decisions to be made with respect to the risks of health and what measures for prevention or control should be taken.

Preventing or minimising the risk

Where a reasonably foreseeable risk has been identified the use of water systems, etc. that could lead to exposure should be avoided, so far as is reasonably practicable, until measures have been taken to minimise the risk. The scheme to minimise the risk should be specific and sufficiently detailed to enable it to be implemented and managed. It must contain information concerning the plant or system necessary to minimise the risk from exposure and ideally should include :

An up-to-date plan, showing the layout of the plant or system

A description of the correct and safe operation of the plant or system

The precautions to be taken.

The precautions to minimise the risk would include the following :

134

Minimisation of the release of water spray

Avoidance of water temperatures and conditions that favour the growth of Legionellae and other micro-organisms

Avoidance of water stagnation

Avoidance of the use of materials that harbour bacteria and other micro-organisms or provide nutrients for microbial growth

Maintenance of a clean system and the water in it

Use of water treatment techniques

Action to ensure correct and safe operation and maintenance.

Any scheme must include control measures and the necessary checks to ensure that the control is effective.

Control

Where assessment has shown there is a reasonable foreseeable risk, the employer, self-employed person or person in control of the premises should appoint a person to take managerial responsibility for the implementation of precautions.

Record keeping

The responsible person must ensure that suitable records are kept, including :

The name and position of the person having managerial responsibility

The assessment of the risk, and the name and position of the person who carried out the assessment.

A written scheme for minimising or eliminating the risk.

The names and positions of the persons responsible for implementing the scheme.

Method of management of the scheme.

The plant systems are in use and their state, for example if they are drained down.

Records should be kept with respect to the implementation of the scheme. The records should show :

The precautionary measures taken with clear indication that they have been correctly carried out, when and by whom

The results of any inspection, test or check carried out, including the dates and the names of the persons carrying out the checks.

A register of remedial works necessary, including the date the work was completed.

Practical implementation

The Health and Safety Executive Publication "The Control of Legionellosis including Legionnaire's Disease" HSG 70 provides practical guidance on the implementation of precautionary and preventive measures.

References

Approved Code of Practice - The prevention or control of Legionellosis (including Legionnaire's Disease) - HSC L8(rev).

The control of Legionellosis, including Legionnaire's Disease HS(G)70 (formerly guidance note EH48) - HSE.

Legionellosis - An interpretation of the requirements of the Health and Safety Commission's Approved Code of Practice : The prevention and control of Legionellosis CIBSE GN3 : 1993.

Appendix A   LEGISLATION, APPROVED CODES OF PRACTICE, AND GUIDANCE FROM HSE

Year   SI No.   Title

1974   **Health and Safety at Work etc. Act 1974**

Approved Code of Practice
COP 26 Rider operated lift trucks - operator training: approved code of practice and supplementary guidance. ISBN 0 7176 0474 8.

Approved Code of Practice
L8 The prevention or control of Legionellosis (including legionnaires disease). ISBN 0 7176 0457 8.

Guidance
L1 A guide to the Health and Safety at Work etc. Act. ISBN 0 7176 0441 1.

1987   2115   The Control of Asbestos at Work Regulations

Approved Code of Practice
L27 The Control of Asbestos at Work. Control of Asbestos at Work Regulations 1987. Approved code of practice. 2nd Edition 1993. ISBN 0 11 882037 0.

Approved Code of Practice
L28 Work with asbestos insulation, asbestos coating and asbestos insulation board. Control of Asbestos at Work Regulations 1987. Approved code of practice. 2nd Edition 1993. ISBN 0 11 882038 9.

1989   0635   The Electricity at Work Regulations

Approved Code of Practice
COP 34 The use of electricity in mines. Electricity at Work Regulations 1989. Approved code of practice 1989. ISBN 0 11 885436.

Approved Code of Practice
COP 35 The use of electricity at Quarries. Electricity at Work Regulations 1989. Approved code of practice 1989. ISBN 0 11 885484 4.

Guidance
HS(R)25 Memorandum of guidance on the Electricity at Work Regulations 1989. ISBN 0 11 883963 2.

Year	SI No.	Title

1992   **2051**   **The Management of Health and Safety at Work Regulations**

Amended by 1994/2865 the Management of Health and Safety at Work (Amendment) Regulations.

Approved Code of Practice
L21 Management of Health and Safety at Work Regulations 1992. Approved code of practice 1992. ISBN 0 7176 0412 8.

1992   **2792**   **The Health and Safety (Display Screen Equipment) Regulations**

Guidance
L26 Display Screen Equipment at Work. Health and Safety (Display Screen Equipment) Regulations 1992. Guidance on the Regulations ISBN 0 7176 0410 1.

Guidance
HS(G)90 VDUs: An easy guide to the Regulations. ISBN 0 7176 0735 6.

1992   **2793**   **The Manual Handling Operations Regulations**

Guidance
L23 Manual Handling Operations Regulations 1992. Guidance on regulations. ISBN 0 7176 0411X.

1992   **3004**   **The Workplace (Health, Safety and Welfare) Regulations**

Approved Code of Practice
L24 Workplace health, safety and welfare.
ISBN 07176 04136.

1998   **The Provision and Use of Work Equipment Regulations**

Guidance
L22 (second edition) Safe Use of work equipment.
Provision and use of work equipment regulations 1998.
Approved Code of Practice and Guidance
ISBN 0 7176 16266

1994   **1866**   **The Gas Safety (Installation and Use) Regulations**
                            Approved Code of Practice and Guidance

L56 Safety in the installation and use of gas systems and appliances. Gas Safety (Installation and Use) Regulations 1994. Approved code of practice and guidance. ISBN 0 7176 0797 6.

Year	SI No.	Title
Year	*SI No.*	*Title*

1994   **1063**   **The Supply of Machinery (Safety) (Amendment) Regulations**

1994   **2865**   **The Management of Health and Safety at Work (Amendment) Regulations**

Guidance
HS(G)122 New and Expectant Mothers at Work. A Guide for Employers.

1994   **3098**   **The Simple Pressure Vessels (Safety) (Amendment) Regulations**

1994   **3140**   **The Construction (Design and Management) Regulations 1994**

Approved Code of Practice
L54 Managing construction for Health and Safety: Construction (Design and Management) Regulations 1994. Approved code of practice 1995. ISBN 07176 0792 5.

1994   **3246**   **The Control of Substances Hazardous to Health Regulations**

Amended by 1994/3247 the Chemicals (Hazard Information and Packaging for Supply) Regulations.

1995   **The Reporting of Injuries, Diseases and Dangerous Occurrences Regulations**

1995   L73 A Guide to the Reporting of Injuries, Diseases and Dangerous Occurrences Regulations 1995 (RIDDOR 95).

Publications are available from:

HSE Books
PO Box 19999
Sudbury
Suffolk
CO10 6FS

Telephone :   01787 881 165
Fax :   01787 313 995

Health and Safety Enquiries

HSE Infoline Telephone: 0541 545500
or write to:
HSE Information Centre,
Broad Lane, Sheffield S3 7HQ.

Appendix B TYPICAL SAFETY INSTRUCTIONS

Part 1 - General Provisions

Scope

These Safety Instructions are for general application for work involving either, or both, non-electrical and electrical work as further described in Parts 2 and 3.

DEFINITIONS

(for use with this safety instruction)

Approved

Sanctioned in writing by the responsible director in order to satisfy in a specified manner the requirements of any or all of these Safety Rules.

Appliance

A device requiring a supply of electricity to make it work.

Company

To be defined.

Responsible Director

The Director of the Company, partner or owner responsible for safety.

Conductor

An electrical conductor arranged to be electrically connected to a system.

Competent Person

Person required to work on electrical equipment, installations and appliances and recognised by the Employer as having sufficient technical knowledge and/or experience to enable him/her to carry out the specified work properly without danger to themselves or others. It is recommended that this competence should be recognised by means of written documentation.

Customer

A person, or body, that has a contractual relationship with the Employer for the provision of goods or services.

Danger

Risk of injury to persons (and livestock where expected to be present) from:

(i)   fire, electric shock and burns arising from the use of electrical energy, and

(ii) mechanical movement of electrically controlled equipment, insofar as such danger is intended to be prevented by electrical emergency switching or by electrical switching for mechanical maintenance of non-electrical parts of such equipment.

Dead

At or about zero voltage in relation to earth, and disconnected from any live system.

Earth

The conductive mass of the Earth, whose electric potential at any point is conventionally taken as zero.

Earthed

Connected to Earth through switchgear with an adequately rated earthing capacity or by approved earthing leads.

Electrical Equipment

Anything used, intended to be used or installed for use to generate, provide, transmit, transform, rectify, convert, conduct, distribute, control, store, measure or use electrical energy.

Electrical Installation

An assembly of associated electrical equipment supplied from a common origin to fulfil a specific purpose and having certain co-ordinated characteristics.

Isolated

The disconnection and separation of the electrical equipment from every source of electrical energy in such a way that this disconnection and separation is secure.

Live

Electrically charged.

Notices

Caution Notice - A notice in approved form conveying a warning against interference.

Danger Notice - A notice in approved form reading "Danger".

Supervisor

(i) Immediate Supervisor - a person (having adequate technical knowledge, experience and competence) who is regularly available at the location where work is in progress or who attends the work area as is necessary to ensure the safe performance and completion of work.

(ii) Personal Supervisor - a person (having adequate technical knowledge, experience and competence) such that he/she is at all times during the course of the work in the presence of the person being supervised.

Voltage

Voltage by which an installation (or part of an installation) is designated. The following ranges of nominal voltage (rms values for a.c.) are defined:

- **Extra-low**. Normally not exceeding 50 V a.c. or 120 V ripple free d.c., whether between conductors or to Earth,
- **Low**. Normally exceeding extra-low but not exceeding 1000 V a.c. or 1500 V d.c. between conductors, or 600 V a.c. or 900 V d.c. between conductors and Earth.

The actual voltage of the installation may differ from the nominal value by a quantity within normal tolerances.

- **High Voltage (HV)**. A voltage exceeding 1000 V a.c. or 1500 V d.c. between conductors, or 600 V a.c. or 900 V d.c. between conductors and Earth.

Basic Requirements

1.1   Other Safety Rules and related procedures

In addition to the application of these Safety Instructions, other rules and procedures as issued by the Employer, or by other authorities, shall be complied with in accordance with management instructions.

In that employees may be required to work in locations, or on or near electrical equipment, installations and appliances, that are not owned or controlled by the Employer, these Safety Instructions have been produced to reasonably ensure safe working, since no other rules/instructions will normally be applicable. However, where the owner has his own rules/instructions and procedures, agreement shall be reached between the Company and the owner on which rules/instructions shall be applied. Such agreement shall be made known to the employees concerned.

1.2 Information and Instruction

Arrangements shall be made to ensure:

(i)   that all employees concerned are adequately informed and instructed as to any equipment, installations or appliances which are associated with work and which legal requirements, Safety Rules and related procedures shall apply; and,

(ii)   that other persons who are not employees but who may be exposed to danger by the work also receive reasonably adequate information.

1.3 Issue of Safety Instructions

Employees and other persons issued with safety instructions shall sign a receipt for a copy of these Safety Instructions (and any amendments thereto) and shall keep them in good condition and have them available for reference as necessary when work is being carried out under these Safety Instructions.

1.4 Special Procedures

Work on, or test of, equipment, installations and appliances to which rules cannot be applied, or for special reasons should not be applied, shall be carried out in accordance with recognised good practice.

1.5 Objections

When any person receives instructions regarding work covered by these Safety Instructions and objects, on safety grounds, to the carrying out of such instructions, the person issuing them shall have the matter investigated and, if necessary, referred to a higher authority for a decision before proceeding.

1.6 Reporting of Accidents and Dangerous Occurrences

All accidents and dangerous occurrences, whether of an electrical nature or not, shall be reported in accordance with The Reporting of Injuries, Diseases and Dangerous Occurrences Regulations 1995.

1.7 Health and Safety

The employer and all employees have a duty to comply with the relevant provisions of the Health and Safety at Work etc. Act 1974 and with other relevant statutory provisions such as the Factories Act 1961 and the various Regulations affecting health and safety, including electrical safety. Additionally, authoritative guidance is available from the Health and Safety Executive and other sources.

In addition to these statutory duties and any other responsibilities separately allocated to them, all persons who may be concerned with work as detailed in Section 1.1 shall be conversant with, and comply with, those Safety Instructions and codes of practice relevant to their duties. Ignorance of legal requirements, or of Safety Instructions and related procedures, shall not be accepted as an excuse for neglect of duty. If any person has any doubt as to any of these duties he should report the matter to his immediate supervisor.

1.8   Compliance with safety instructions

It is the duty of everyone who may be concerned with work covered by these Safety Instructions, to ensure their implementation and to comply with them and related codes of practice. Ignorance of the relevant legal requirements, Safety Instructions, Codes of Practice or approved procedures is not an acceptable excuse for neglect of duty.

The responsibilities placed upon persons may include all or part of those detailed in this section, depending on the role of the persons.

Any written authorisation given to persons to perform their designated role in implementing the Safety Instructions must indicate the work permitted.

Whether employees are authorised as competent or not, all have the following duties which they must ensure are implemented:

-        All employees shall comply with these Safety Instructions when carrying out work, whether instructions are issued orally or in writing.

-        All employees shall use safe methods of work, safe means of access and the personal protective equipment and clothing provided for their safety.

-        All employees when in receipt of work instructions shall:

(i)   be fully conversant with the nature and extent of the work to be done;
(ii)   read the contents and confirm to the person issuing the instructions that they are fully understood;
(iii)   during the course of the work, adhere to, and instruct others under their charge to adhere to, any conditions, instructions or limits specified in the work instructions;
(iv)   when in charge of work, provide immediate or personal supervision as required.

Part 2 - Non-electrical

Scope

The non-electrical part of the Safety Instructions shall be applied to work by employees, in the activities that are non-electrical. This work may involve:-

(i) Work on customers' premises
(ii) Work on employer's premises
(iii) Work on the public highway or in other public places.

The Safety Instructions applicable to this work are those contained in Parts 1 and 2 of the Safety Instructions. When work of an electrical nature is being carried out, all Parts (1, 2 and 3) of the Safety Instructions apply.

Basic Safety Precautions

2.1 General Principle

The general principle is to avoid accidents. Most accidents arise from simple causes and can be prevented by taking care.

2.2 Protective Clothing and Equipment

The wearing of protective clothing and the use of protective equipment can, in appropriate circumstances, considerably reduce the severity of injury should an accident occur.

Where any work under these Safety Instructions and related procedures takes place, appropriate safety equipment and protective clothing of an approved type shall be issued and used.

At all times employees are expected to wear sensible clothing and footwear having regard to the work being carried out. Further references are made, in particular circumstances, to the use of gloves. Hard hats must be worn at all times when there is a risk of head injury and particularly on building sites.

Where there is danger from flying particles of metal, concrete or stone, suitable eye protection must be provided and must be used by employees. If necessary, additional screens must be provided to protect other persons in the vicinity.

2.3 Good Housekeeping

Tidiness, wherever work is carried out, is the foundation of safety; good housekeeping will help to ensure a clean, tidy and safe place of work.

Particular attention should be paid to:-

(i) picking up dropped articles immediately;

(ii) wiping up any patches of oil, grease or water as soon as they appear and if necessary, spreading sand or sawdust;

(iii) removing rubbish and scrap to the appropriate place;

(iv) preventing objects falling from a height by using containers for hand tools and other loose material;

(v)   ensuring stairs and exits are kept clear.

No job is completed until all loose gear, tools and materials have been cleared away and the workplace left clean and tidy. Most falls are caused by slippery substances or loose objects on the floor and good housekeeping will remove most of the hazards that can occur.

2.4   Safe Access

It is essential that every place of work is at all times provided with safe means of access and exit, and these routes must be maintained in a safe condition.

Keeping the workplace tidy minimises the risk of falling which is the major cause of accidents, but certain special hazards associated with work in confined spaces require particular attention.

2.4.1   Ladders

All ladders should be of sound construction, uniquely identified and free from apparent defects. This is of particular importance in connection with timber ladders. The following practices should always be observed:-

Ladders should be checked before use. Any defects must be reported and the equipment clearly marked and not used until repaired.

All ladders should be regularly inspected by a competent person and a record kept.

Ladders in use should stand on a level and firm footing. Loose packing should not be used to support the base.

Ladders should be used at the correct angle, i.e., for every four metres up, the bottom of the ladder must be one metre out.

Ladders should be lashed at the top when in use, but when this is not practicable they should be held secure at the bottom.

The ladder top should extend to a height of at least one metre above any landing place.

Hand tools and other material should not be carried in the hand when ascending or descending ladders. A bag and sash line should be used.

Suitable crawling ladders or boards must be used when working on asbestos cement and other fragile roofs. Permanent warning notices should be placed at the means of access to these roofs.

(**NB:** In a situation where no ladder is available, and the work requires a small step up, it is the employee's responsibility to ensure that any other article used for the purpose is totally suitable.)

2.4.2   Openings in Floors

Every floor opening must be guarded, and it is important that other occupants of the workplace are made aware of these hazards.

In addition to the risk of persons falling through any opening, there is also a risk from falling objects and safe placing of tools, materials and other objects when working near openings, holes or edges, or at any height, will also prevent accidents.

If work has to be carried out in confined spaces such as tunnels and underground chambers, the atmosphere may be deficient in oxygen or may contain dangerous fumes or substances. The Electricity Association Engineering Recommendations, ERG64, Safety in Cableways must be followed in these circumstances.

2.5   Lighting

Good lighting, whether natural or artificial, is essential to the safety of people whether at the workplace or moving about. If natural lighting is inadequate, it must be supplemented by adequate and suitable artificial lighting. If danger may arise from a power failure, adequate emergency lighting is required.

2.6   Lifting and handling

All employees must be trained in the appropriate lifting and handling techniques according to the type of work undertaken.

2.7   Fire Precautions

All employees must be thoroughly conversant with the procedure to be followed in the event of fire. Whether working on customers' premises or elsewhere, employees should familiarise themselves with escape routes, fire precautions, etc., before commencing work.

Fire exits must always be kept clear, and access to fire fighting equipment unobstructed.

All fire fighting equipment that is the Employer's responsibility must be regularly inspected, maintained and recorded whether by local supervisor, safety supervisor or appropriate third party. Individuals should report any apparent damage to equipment.

2.8   Hand Tools

Hand tools must be suitable for the purpose for which they are being used and are the responsibility of those using them. They must be maintained in good order and any which are worn or otherwise defective must be reported to a supervisor. Approved insulated tools are available for work on live electrical equipment.

2.9   Mechanical Handling

Fork-lift and similar trucks must only be driven by operators who have been properly trained, tested and certified for the type of trucks they have to operate.

Supervisors should control the issue and return of the truck keys and they should ensure that a daily check of the truck and its controls is carried out by the operator.

2.10   Portable Power Tools

All portable electrical apparatus including cables, portable transformers and other ancillary equipment should be inspected before use and maintained and tested at regular intervals.

Trailing cables are frequently damaged and exposed to wet conditions. Users must report all such damage and other defects as soon as possible, and the faulty equipment must be immediately withdrawn from use.

When not in use power tools should be switched off and disconnected from the source of supply.

2.11   Welding, burning and heating processes

2.11.1 General

Welding, burning and heating processes involving the use of gas and electricity demand a high degree of skill and detailed knowledge of the appropriate safety requirements.

Specific safety instructions will be issued to employees using such equipment.

Suitable precautions should be taken, particularly when working overhead, to prevent fire or other injury from falling or flying sparks.

All heating, burning and welding equipment must be regularly inspected, and a record kept.

2.11.2 Propane

Propane is a liquefied petroleum gas stored under pressure in cylinders which must be stored vertically in cool, well-ventilated areas, away from combustible material, heat sources and corrosive conditions. Cylinders must be handled carefully and not allowed to fall from a height; when transported, they must always be carried in an approved restraint.

When the cylinder valve is opened the liquid boils, giving off a highly flammable gas. The gas is heavier than air and can give rise to a highly explosive mixture. It is essential, therefore, that valves are turned off after use.

2.12   Machinery

All machinery will be guarded as necessary to prevent mechanical hazards.

Facilities shall be provided for isolating and locking off the power to machinery. Work on machines shall not commence unless isolation and locking off from all sources of power has been effected and permits issued to work. (See Section 3, in particular 3.3).

Part 3 - Electrical

Scope

This part of the Safety Instructions shall be applied to electrical work. This work may involve:-

(i)   Work on employers equipment;
(ii)   Work on customers' electrical installations;
(iii) Work on customers' electrical appliances.

This work will normally be concerned with equipment, installations and appliances at low voltages. In the event of work needing to be carried out on high voltage equipment and installations (i.e. where the voltage exceeds 1000 volts a.c.), additional instructions and procedures laid down for High Voltage work must be issued to those employees who carry out this work.

Basic Safety Precautions

3.1   General Principle

As a general principle, and wherever reasonably practicable, work should only be carried out on equipment that is dead and isolated from all sources of supply. Such equipment should be proved dead by means of an approved voltage testing device which should be tested before and after verification, or by clear evidence of isolation taking account of the possibility of wrong identification or circuit labelling.

Equipment should always be assumed to be live until it is proved dead. This is particularly important where there is a possibility of backfeed from another source of supply.

3.2   Information Prior to Commencement of Work

According to the complexity of the installation, the following information may need to be provided before specified work is carried out:-

(i)   details of the supply to the premises, and to the system and equipment on which work is to be carried out;

(ii)   details of the relevant circuits and equipment and the means of isolation;

(iii)   details of any customer's safety rules or procedures that may be applicable to the work;

(iv)   The nature of any processes or substances which could give rise to a hazard associated with the work, or other special conditions that could affect the working area, such as the need for special access arrangements;

(v)   emergency arrangements on site;

(vi)   the name and designation of the person nominated to ensure effective liaison during the course of the work.

Where the available information, or the action to be taken as a result of it, is considered by the person in charge of the work to be inadequate for safe working, such work should not proceed until that inadequacy has been removed or a decision obtained from a person in higher authority. Defects affecting safe working should be reported to the appropriate supervisor.

3.3   Precautions to be taken before Work Commences on Dead Electrical Equipment

In addition to any special precautions to be taken at the site of the work, such as for the presence of hazardous processes or substances, the following electrical precautions should be taken, according to the circumstances, before work commences on dead electrical equipment:

(i)   the electrical equipment should first be properly identified and disconnected from all points of supply by the opening of circuit-breakers, isolating switches, the removal of fuses, links or current-limiting devices, or other suitable means. Approved Notices, warning against interference, should be affixed at all points of disconnection.

(ii)   all reasonably practicable steps should be taken to prevent the electrical equipment being made live inadvertently. This may be achieved, according to the circumstances, by taking one or more of the following precautions:

(a)   approved locks should be used to lock off all switches etc. at points where the electrical equipment and associated circuits can be made live. This should be additional to any lock applied by any other party; the keys to all locks should be retained by the person in charge of the work or in a specially provided key safe,

(b) any fuses, links or current-limiting devices involved in the isolation procedures should be retained in the possession of the person in charge of the work, and

(c) in the case of portable apparatus, where isolation has been by removal of a plug from a socket-outlet, suitable arrangements should be made to prevent unauthorised re-connection,

(d) approved notices should be placed at points where the electrical equipment and associated circuits can be made live.

(iii) the electrical equipment should be proved dead by the proper use of an approved voltage testing device and/or by clear evidence of isolation, such as physically tracing a circuit. Approved testing devices should be checked immediately before and after use to ensure that they are in working order,

(iv) when work is carried out on timeswitched or other automatically controlled equipment or circuits, the fuses or other means of isolation controlling such equipment or circuits should be removed. On no account should reliance be placed on the timeswitches, limit switches, lock-out push buttons etc., or on any other auxiliary equipment, as means of isolation,

(v) where necessary, approved notices should be displayed to indicate any exposed live conductors in the working zone,

(vi) when it is required to work on dead equipment situated in a substation or similar place where there are exposed live conductors, or adjacent to High Voltage plant, the safe working area should be defined by a person authorised in writing under the Safety Rules or under procedures controlling that plant, and all subsequent work must be conducted in accordance with such rules or procedures. Where necessary the exposed live conductors should be adequately screened in an approved manner or other approved means taken to avoid danger from the live conductors.

3.4 Precautions to be taken before Work Commences on or near Live Equipment

No person shall be engaged in any work activity on or so near any live conductor (other than one suitably covered with insulating material so as to prevent danger) that danger may arise unless-

(i) it is unreasonable in all the circumstances for it to be dead; and

(ii)   it is reasonable in all the circumstances for them to be at work on or near it while it is live; and

(iii)   suitable precautions (including where necessary the provision of suitable protective equipment) are taken to prevent injury. (Regulation 14, Electricity at Work Regulations).

Where work is to be carried out on live equipment the following protective equipment should be provided, maintained and used, by adequately trained personnel, in accordance with the Safety Rules or local procedures as appropriate:

(i)   approved screens or screening material,

(ii)   approved insulating standards in the form of hardwood grating or approved rubber insulating mats,

(iii)   approved insulated tools, and

(iv)   approved insulating gloves.

When testing, including functional testing or adjustment of electrical equipment, requires covers to be removed so that terminals or connections that are live, or can be made live, are exposed, precautions should be taken to prevent unauthorised approach to or interference with live parts. This may be achieved by keeping the work area under the immediate surveillance of an employee or by erecting a suitable barrier, with Approved Notices displayed warning against approach and interference. When live terminals or site barriers are being adjusted, only approved insulated tools should be used.

Additional precautions may be required because of the nature of any hazardous process or special circumstances present at the site of the work.

Work on live equipment should only be undertaken where it is unreasonable in all the circumstances for it to be made dead.

3.5   Operation of Switchgear

The operation of switchgear should only be carried out by a Competent Person after he has obtained full knowledge and details of the installation and the effects of the intended switching operations.

Under no circumstances must equipment be made operable by hand signals or by a pre-arranged time interval.

Figure B3.1 Model form of permit-to-work

MODEL FORM OF PERMIT-TO-WORK (FRONT)

PERMIT-TO-WORK

1. ISSUE   No ....

To ....

The following apparatus has been made safe in accordance with the Safety Rules for the work detailed on this Permit-to-Work to proceed:

....

....

TREAT ALL OTHER APPARATUS AS LIVE

Circuit Main Earths are applied at

....

....

Other precautions and information required and any local instructions applicable to the work, notes 1 and 2.

....

....

The following work is to be carried out :...

....

....

Name (Block capitals) ....

Signature...

         Time ....   Date ....

5.   Diagram

The diagram should show :

(a)   the safe zone where work is to be carried out

(b)   the points of isolation

(c)   the places where earths have been applied, and

(d)   the locations where 'danger' notices have been posted.

MODEL FORM OF PERMIT-TO-WORK (BACK)

2. RECEIPT
(note 2)

I accept responsibility for carrying out the work on the Apparatus detailed on this Permit-to-Work and no attempt will be made by me, or by the persons under my charge, to work on any other Apparatus.

Name (Block capitals)...

Signature ....

    Time...   Date..

3. CLEARANCE
(note 3)

All persons under my charge have been withdrawn and warned that it is no longer safe to work on the Apparatus detailed on this Permit-to-Work, and all Additional Earths have been removed.

The work is complete* / incomplete*

All gear and tools have* / have not* been removed

Name (Block capitals)...

Signature ....

    Time ....   Date..

*Delete words not applicable

4. CANCELLATION
(note 3)

This Permit-to-Work is cancelled.

Name (Block capitals)...

Signature ....

    Time...   Date..

5

Notes on Model Form of Permit-to-Work

1   ACCESS TO AND WORK IN FIRE PROTECTED AREAS

Automatic control

Unless alternative **Approved** procedures apply because of special circumstances then before access to, or work or other activities are carried out in, any enclosure protected by automatic fire extinguishing equipment:

(a)   The automatic control shall be rendered inoperative and the equipment left on hand control. A **Caution Notice** shall be attached.

(b)   Precuations taken to render the automatic control inoperative and the conditions under which it may be restored shall be noted on any **Safety Document** or written instruction issued for access, work or other activity in the protected enclosure.

(c)   The automatic control shall be restored immediately after the persons engaged on the work or other activity have withdrawn from the protected enclosure.

2.   PROCEDURE FOR ISSUE AND RECEIPT

(a)   A **Permit-to-Work** shall be explained and issued to the person in direct charge of the work, who after reading its contents to the person issuing it, and confirming that he understands it and is conversant with the nature and extent of the work to be done, shall sign its receipt and its duplicate.

(b)   The recipient of a **Permit-to-Work** shall be a **Competent Person** who shall retain the **Permit-to-Work** in his possession at all times whilst work is being carried out.

(c)   Where more than one **Working Party** is involved a **Permit-to-Work** shall be issued to the **Competent Person** in direct charge of each **Working Party** and these shall, where necessary, be cross-referenced one with another.

3.   PROCEDURE FOR CLEARANCE AND CANCELLATION

(a)   A **Permit-to-Work** shall be cleared and cancelled:

    (i)   when work on the **Apparatus** or **Conductor** for which it was issued has been completed;

    (ii)   when it is necessary to change the person in charge of the work detailed on the **Permit-to-Work**;

    (iii)   at the discretion of **Responsible Person** when it is necessary to interrupt or suspend the work detailed on the **Permit-to-Work**.

(b)   The recipient shall sign the clearance and return to the **Responsible Person** who shall cancel it. In all cases the recipient shall indicate in the clearance section whether the work is "complete" or "incomplete" and that all gear and tools "have" or "have not" been removed.

(c)   Where more than one **Permit-to-Work** has been issued for work on **Apparatus** or **Conductors** associated with the same **Circuit Main Earths**, the **Controlling Engineer** shall ensure that all such **Permits-to-Work** have been cancelled before the **Circuit Main Earths** are removed.

4.   PROCEDURE FOR TEMPORARY WITHDRAWAL OR SUSPENSION

Where there is a requirement for a **Permit-to-Work** to be temporarily withdrawn or suspended this shall be in accordance with an **Approved** procedure.

Appendix C   REPORTING OF ACCIDENTS

C.1   Introduction

The Health and Safety Executive publish a guide to The Reporting of Injuries, Diseases And Dangerous Occurrences Regulations 1995, reference L 73.

The following events are reportable within an appropriate period of time :

When someone at work, is unable to do their normal work for more than 3 days, as a result of an injury caused by an accident at work;

On the death of an employee, if this occurs sometime after a reportable injury which led to that employee's death, but not more than 1 year afterwards;

A person at work suffers one of a number of specified diseases. Specified diseases are set out in a schedule to the Regulations.

Reporting and notifying

In the event of there being a notifiable event the responsible person shall :

1.   Forthwith notify the relevant enforcing authority thereof by the quickest practical measure, and within 10 days send a report thereof to the relevant enforcing authority on a form approved for the purpose.

The responsible person in general is the person at the time having control of the premises at which, or in connection with the work of which, the accident or dangerous occurrence or reportable disease happened.

If you have an accident on your premises, you may not be guilty of any offence but you certainly will be if you do not report it. The Reporting of Injuries, Diseases And Dangerous Occurrences Regulations 1995 came into force on 1 April 1996. They apply a single set of report requirements to all work activities in Great Britain and in the off-shore oil and gas industries.

The reports will be compiled by the Health and Safety Executive and Local Authorities. This should provide valuable information as to where the risks are, how they arise, and indicate if there are any unfavourable trends. The data can also be useful to guide employers as to how best they might prevent injury and ill health.

C.2   Where the Regulations apply

The Regulations apply in places of work, and to events that arise out of or in connection with work activities, as covered by the Health and Safety at Work etc. Act. The Regulations apply to Great Britain but not Northern Ireland, where separate Regulations are to be made. They also apply to certain work activities carried out in United Kingdom territorial waters. The Regulations apply to mines under the sea and other activities in territorial waters, such as loading and unloading, construction, repair and diving.

C.3   At the accident

If any of the following accidents should occur, the enforcing authority must be notified by the quickest practical means, such as by telephone :

a)   the death of any person as a result of an accident, whether or not they are at work;
b)   a person suffers a major injury as a result of an accident;
c)   someone who is not at work suffers an injury as a result of an accident arising out of or in connection with work and that person is taken to hospital.

If there is a dangerous occurrence

Dangerous occurrences are accidents or events which have not necessarily resulted in a reportable injury, but have the potential to cause such an injury.

Dangerous occurrences are listed in schedule 2 to the Regulations. As examples of the sort of dangerous occurrences that are reportable and are concerned with electrical matters, the following are included :

Electrical short-circuit

Electrical short-circuit or overload attended by fire or explosion which results in the stoppage of the plant involved for more than 24 hours, or which has the potential to cause the death of any person.

Overhead lines

Any unintentional incident in which plant or equipment either:

Comes into contact with an uninsulated overhead electric line of which the voltage exceeds 200 volts, or

Causes an electrical discharge from such an electric line by coming into close proximity with it.

Pressure systems

The failure of any closed vessel (including a boiler or tube boiler) or any associated pipework, in which the internal pressure was above or below atmospheric pressure, where the failure has the potential to cause the death of any person.

Lifting Machinery etc.

The collapse or overturning of, or failure of any load-bearing part of any lift or hoist, crane or derrick, mobile power access platform, access cradle or window cleaning cradle, excavator, pile driving frame or fork-lift trunk.

The above are just samples of the report occurrences.

C.4   Enforcing authority

The enforcing authority may be either the Health and Safety Executive or the Local Authority. The exact split is determined by the Health and Safety (Enforcing Authority) Regulations 1989. However, Local Authorities are generally responsible for enforcing Health and Safety legislation in:

retailing
some warehouses
most offices
hotels and catering
sports
leisure
consumer service
places of worship.

If in doubt as to whether there is a need to report, or as to whether it should be reported to the Health and Safety Executive or the Local Authority, the Health and Safety Executive can be immediately telephoned.

The report forms are available from the Health and Safety Executive. Typical report form is provided in Figures C.1.

C.5   Keeping of records

The responsible person is required to keep a record of:

1.   Any event which is required to be reported upon.
2.   Any disease required to be reported upon.
3.   Any such particulars as may be approved by the Health and Safety Executive.

The records are required to be kept either at the place of work to which they relate, or at the usual place of business of the responsible person.

Extracts from records must be sent to the Enforcing Authority on request and, additionally, an inspector from the Enforcing Authority may require any part of the records to be produced.

Figure C.1 The report form

Health and Safety at Work etc. Act 1974

The Reporting of Injuries, Diseases and Dangerous Occurrences Regulations 1995

Report of an injury of dangerous occurrence

Filling in this form

This form must be filled in by an employer or other responsible person.

Part A

About you

1. What is your full name?

2. What is your job title?

3. What is your telephone number?

About your organisation

4. What is the name of your organisation

5. What is its address and postcode?

6. What type of work does the organisation do?

Part B

About the incident

1. On what date did the incident happen?

/   /

2. At what time did the incident happen?
(Please use the 24-hour clock eg 0600)

3. Did the incident happen at the above address?
Yes ☐ Go to question 4
No ☐ Where did the incident happen?
☐ elsewhere in your organisation - give the name, address and postcode
☐ at someone else's premises - give the name, address and postcode at in a
☐ public place - give details of where it happened

If you do not know the postcode, what is the name of the local authority?

4. In which department, or where on the premises, did the incident happen?

Part C

About the injured person

If you are reporting a dangerous occurrence, go to Part F.

If more than one person was injured in the same accident, please attach the details asked for in Part C and Part D for each injured person.

1. What is their full name?

2. What is their home address and postcode?

3. What is their home phone number?

4. How old are they?

5. ☐ Are they
☐ male?
☐ female?

6. What is their job title? ☐☐☐

7. Was the injured person (tick only one box)
one of your employees?
on a training scheme? Give details:

on work experience?
employed by someone else? Give details of the employer:

☐ self-employed and at work?
☐ a member of the public?

Part D

About the injury

1. What was the injury? (e.g. fracture, laceration) ☐☐

2. What part of the body was injured? ☐☐

Continued overleaf

3. Was the injury (tick the box that applies)

☐ a fatality?

☐ a majory injury or condition (see accompanying notes)

☐ an injury to employee or self-employed person which prevented them doing their normal work for more than 3 days?

☐ an injury to a member of the public which meant they had to be taken from the scene of the accident to a hospital for treatment?

4. Did the injured person (tick all boxes that apply)

☐ become unconscious?

☐ need resuscitation?

☐ remain in hospital for more than 24 hours?

☐ none of the above.

Part E

About the kind of accident

Please tick the one box that best describes what happened, then go to Part G.

☐ Contact with moving machinery or material being machined

☐ Hit by moving, flying or falling object

☐ Hit by a moving vehicle

☐ Hit something fixed or stationary

☐ Injured while handling, lifting or carrying

☐ Slipped, tripped or fell on the same level

☐ Fell from a height

How high was the fall?

metres

☐ Trapped by something collapsing

☐ Drowned or asphyxiated

☐ Exposed to, or in contact with, a harmful substance

☐ Exposed to fire

☐ Exposed to an explosion

☐ Contact with electricity or an electrical discharge

☐ Injured by an animal

☐ Physically assaulted by a person

☐ Another kind of accident (describe it in Part G)

Part F

Dangerous occurrences

Enter the number of the dangerous occurrence you are reporting. (The numbers are given in the Regulations and in the notes which accompany this form.)

Part G

Describing what happened

Give as much detail as you can. For instance

- the name of any substance involved
- the name and type of any machine involved

- the events that led to the accident
- the part played by any people.

If it was a personal injury, give details of what the person was doing. Describe any action that has since been taken to prevent a similar accident. Use a separate piece of paper if you need to.

Part H

Your signature

Signature

Date

/	/

Where to send the form

Please send it to the Enforcing Authority for the place where it happened. If you do not know the Enforcing Authority, send it to the nearest HSE office.

For official use			
Client number	Location number	Event number	
			INV ☐ REP ☐ Y ☐ N

INDEX